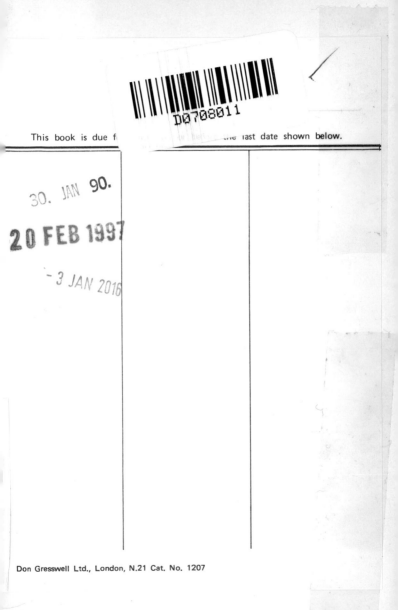

This book is due f ... ... ~~e last date shown below.

Don Gresswell Ltd., London, N.21 Cat. No. 1207

PELICAN BOOKS

THE CHEMISTRY OF LIFE

~~~~~~~~~~~~ 1928 in London. He went
~~~~~~~~~~~~~~~~~~~~~~ ⸴llege, Cambridge.
After ⸴⸴~~~~~~~~~~~~~~~~~~~~~⸴ y he read for his
doctorate at the Maudsley Institute of Psychiatry, London. He
held both the Beit Memorial and Guinness Fellowships at
New College, Oxford, and was then for five years at Imperial
College, London. He is now Professor of Biology at the Open
University.

The main theme of his research has been the investigation of
the biochemical basis of brain function and memory and the
problem of control processes within the living cell, on which he
has written more than a hundred and fifty research papers.

His books include *The Conscious Brain* (Penguin, 1976) and,
with Hilary Rose, a sociologist, *Science and Society* (Allen
Lane, 1969), *The Political Economy of Science* and *The Radi-
calisation of Science* (1976).

Cath Sanderson graduated from Newcastle upon Tyne in
microbiology in 1972. She has worked in dental research at the
University of Leeds, and was a research assistant in the
Department of Physiology at Newcastle before joining the
Brain Research Group at the Open University, where she is
studying the development of neurotransmitter systems in the
rat brain.

STEVEN ROSE
WITH
CATH SANDERSON

THE CHEMISTRY
OF LIFE

Second Edition

PENGUIN BOOKS

Penguin Books Ltd, Harmondsworth, Middlesex, England
Penguin Books, 40 West 23rd Street, New York, New York 10010, U.S.A.
Penguin Books Australia Ltd, Ringwood, Victoria, Australia
Penguin Books Canada Ltd, 2801 John Street, Markham, Ontario, Canada L3R 1B4
Penguin Books (N.Z.) Ltd, 182–190 Wairau Road, Auckland 10, New Zealand

—

First published 1966
Reprinted 1968
Reprinted with revisions 1970
Reprinted 1971, 1972, 1974, 1975, 1976, 1977
Second edition 1979
Reprinted 1980, 1982, 1983

—

—

Made and printed in Great Britain
by Hazell Watson & Viney Ltd,
Aylesbury, Bucks
Set in Monotype Times Roman

CONTENTS

DEDICATION

*For Hilary, who insisted that biochemistry
be made intelligible*

PREFACE TO THE
SECOND EDITION (1979)

The first edition of *The Chemistry of Life* was published in 1966. I had only a few years since completed a 'classical' biochemical training at Cambridge, done a Ph.D in brain biochemistry in London and was just finishing off a postdoctoral period in Rome when I wrote the book in 1963–4. I was full of optimism at the time, a biochemical positivism which reflected the exuberant atmosphere of the advances in biochemistry of the '50s and '60s. Biochemical knowledge might not change the world, I thought, but it would help explain it, and I believed then, as I believe now, in the importance of opening up science and making it accessible to those outside the arcane elitism with which it surrounds itself. Today, I am more conscious of the problematic nature of the biochemical reductionism which the book exuded, and I find the relationship between my science and the social and philosophical framework within which it is embedded more complex than in the past.

I never intended *The Chemistry of Life* as a text book, but merely to convey some of the excitement of biochemistry to a lay reader, by discussing its central concepts with a minimum of sheer fact. However, perhaps by default of alternatives, the book soon became used in schools and university and college ancillary biochemistry and first-year courses, and this use has become extended since the Open University adopted it as a set text for their Science Foundation Course. This left me in a dilemma, as I was aware that the explosive growth of biochemistry was rendering some areas in particular in urgent need of revision, and the direction of my own thoughts and research had led me to re-analyse some of the orientations that I had taken more easily for granted in the early 1960s. In the event, I have waited to take advantage of the fact that the Open University's Science Foundation Course was to be rewritten so as to have the freedom to reconstruct the book without creating problems for students who were using the earlier version.

7

Preface to the Second Edition (1979)

I have given a lot of thought to the question of how far I should try to move the book in the direction of my more recent interests, but this would, I decided, have been a very different venture. Instead, I have tried to retain as much of the earlier structure and mode of analysis as possible – the central themes of biochemistry as analysis, as metabolism and as control – but I have tried to bring them further into line with present-day thinking. This book, then, is still about the internal content of biochemistry, not cell biology, neurobiology or the social structure of the discipline. Its structure and organization has, I hope, benefited from the experience of the intervening years, and the continued feedback from several generations of Open University students and my colleagues in the Biology Department at the Open University. I could never have undertaken the task without being able to find someone to work with, and I was exceptionally fortunate in Cath Sanderson, a person whose vitality, enthusiasm, dedication and determination to argue for her interpretation of the world has been an essential ingredient of this new edition.

Dr Anna Furth and Dr Irene Ridge, of the Biology Department at the Open University, kindly read and commented on sections of the revised text for me and I should also like to thank Peter Wright, at Penguin, for his ready acceptance of the need to prepare a thorough revision of the book, as well as Nancie Pugh for her impeccable translation of hieroglyphs into typed English. None but me, however, should be blamed for any inaccuracies or idiosyncrasies that may remain.

Now, as to the structure of the new edition. Essentially, it follows the old in that Chapter 1 introduces the basic chemical terms to those who are unfamiliar with them, and Chapters 2 and 3 deal respectively with the properties of the small and giant molecules of which the cell is composed. In the earlier edition, the structure of the cell itself was not discussed until much later; this time, and because of the importance we attach to showing some subcellular organelles as composed of higher order hierarchies of macromolecules, the account of the cell forms Chapter 4. This concludes the section on biochemistry as analysis, and we turn in Chapters 5 and 6 to work, enzymes and metabolism. The core of biochemical energetics and metabolism is in Chapters 7 and 8, and it is these

Preface to the Second Edition (1979)

which may be heavy going for the less dedicated. Connoisseurs will note that SI units and chemiosmosis are 'in', 'energy-rich' bonds and the classical squiggle notation 'out'. As in the earlier editions, a bit of judicious skipping would be acceptable here for the less involved. Chapters 9 and 10 deal with biosynthesis, and because of the expansion in theoretical significance and sheer factual information now available on protein and nucleic acid synthesis, these form a chapter of their own. Like Chapter 10, Chapter 11 on control mechanisms has been extensively rewritten to account for new knowledge, especially on hormone action and receptors, but I have rigorously curbed my desire to enlarge on 'The Cell in Action' (Chapter 12) and 'The Unity of Biochemistry' (Chapter 13).

Perhaps because biochemistry has expanded and diversified so much in the period since the early 1960s, I should spell out yet again that the emphasis in the book is throughout on animal *biochemistry*, not cell or molecular biology. Lest this bring down upon my head the wrath of microbiologists and plant biochemists, let me apologize in advance for this deliberate bias; it reflects my own special enthusiasms, and implies no denigration of their favoured organisms.

August 1978   STEVEN ROSE

Acknowledgements

The following acknowledgements are made for the use of photographs: to Dr M. G. Stewart, Ms P. Mullins and Mr D. Spears for Plates 1, 3, 6 and 7; to Dr J. A. Armstrong for Plate 2; to Dr H. Beaufay for Plate 4; to Dr H. Fernandez-Moran for Plate 8a; to Dr H. E. Huxley for Plate 5; and to Dr E. Kellenberger for Plate 8b. Dr Stewart, Ms Mullins and Mr Spears are members of the Brain Research Group of the Open University.

INTRODUCTION

WHAT IS BIOCHEMISTRY?

'Biochemistry is the study of the chemical constituents of living matter and of their functions and transformations during life processes' – the definition of biochemistry in almost any standard textbook.

One does not normally regard a kitchen as having much in common with a laboratory. Most sciences are associated with heavy and elaborate machinery, the creation of high temperatures, vast pressures, or the concentration of great amounts of energy. The prototype of biochemical experiments, however, is the frying of an egg. Material from a biological source (the chicken) is taken and removed from extraneous surrounding substances (the shell), care being taken all the while not to disrupt its natural organized structure (by breaking the yolk). This partially purified biological material is then subjected, under carefully controlled and regulated conditions, to a number of mild chemical and physical treatments (the egg is gently heated in fat, and pepper and salt are added). Whether the ultimate product is fit for anything more than the dustbin depends entirely on the skill with which these separate operations are performed; the margin between a good breakfast and a charred and tasteless mess is very small.

This analogy is not as unlikely as might at first be imagined; it is very probable that most good biochemists would also make good cooks, for, like cooks, they deal with fragile substances derived from living or recently dead animals and plants, which they must handle rapidly, gently and subtly in order to obtain meaningful results. As in cooking, the type of operation performed is all-important; no one would mistake a fried for a boiled or a scrambled egg; the differences in procedure by which one is arrived at rather than another are small (heating in or out of the shell, with or without prior mixing), but they are critical to the end product. So with biochemistry. Small variations in experimental procedure, in the concentration of a reactant, in the acidity or alkalinity of the medium in which the reaction is being

11

carried out, or in its temperature, cause critical changes to occur in the behaviour of the highly complex chemical entities being studied. Whether these behavioural alterations make sense or nonsense will depend on both the experimental and the theoretical ingenuity of the biochemist.

It is for this reason that biochemistry is so recent a science. The wealth of sophisticated techniques and concepts needed to make possible an approach to the understanding of the chemical order and functioning of the living cell is such that they could not have arisen except on the secure foundations of more elementary sciences. In essence, the medieval doctors speculating on the composition of blood, or devotedly distilling retorts filled with urine, were performing biochemical operations. So was the Italian abbot, Lazzaro Spallanzani, who, in 1783, fed hawks with pieces of meat enclosed in wire boxes, trained the birds to vomit up the boxes at various subsequent times, and observed that the action of the gastric juices on the meat was progressively to liquefy it. He deduced that the liquefaction was the result of chemical interaction between certain substances in the gastric juices and the meat; such active principles were later, under the name of *enzymes*, recognized as the cornerstone of modern biochemistry.

Equally a biochemist was Friedrich Wöhler, who, in 1828, synthesized the biological material urea from the non-biological cyanic acid and ammonia, thereby settling a debate as old as alchemy itself, by providing the first unassailable evidence that the substances present in living organisms are chemical entities which differ from those in the chemist's reagent bottles only in their complexity, and not by the introduction of any mysterious hypothesis of the 'nature of life'.

These were pioneer steps indeed, and great strides forward were made in the hundred years between Wöhler's synthesis of urea and the isolation of the first crystalline enzyme – also, by some strange chance, one related to urea, 'urease' – by Sumner in America in 1926. But the really explosive growth of biochemistry has had to wait on the consolidation of chemical theory, and the pushing forward of the frontiers of biology to a region where the distinction between it and 'chemical physiology' became obscure. By the

1930s, the time was at last ripe for the biochemists to take over. The first signs of the new biochemistry emanated from the German laboratories of Meyerhof and Warburg, which in the 1920s and early 1930s housed some of the most brilliant biochemists in the world. With the coming of the Nazis, many of these youngsters fled to England and America. They found refuge in the ever-hospitable laboratories of Frederick Gowland Hopkins at Cambridge, and in such American laboratories as that of the Rockefeller Institute in New York. From this period, and until the late 1960s, dates the dominance of British and American biochemistry.

From then on, the science of biochemistry has expanded at an immense rate, increasing more rapidly even than that of nuclear physics, itself no sluggard. It is still in the full flood of its growth; dramatic breakthroughs in hitherto little-understood areas are becoming so frequent as to be accepted by the practising biochemist almost with weary resignation, whilst, to cope with the tidal wave of learned papers and reports now appearing, publishers are being forced to produce weekly issues of journals which only a short while ago were monthlies or even quarterlies. In order to ensure rapid distribution of their findings, some researchers have now abandoned the journals and instead circulate duplicated copies of their results to a selected mailing list. Not content with this, biochemistry spills over into related disciplines, and biochemists have staged take-over bids for the journals of physiology and chemistry, to say nothing of such general magazines as *Nature* in Britain and *Science* in America. Concurrently with its expansion, biochemistry has itself begun to fragment, so that its practitioners refer to themselves as enzymologists, molecular biologists, chemical microbiologists, neurochemists, even mitochondriologists. To remain a plain biochemist is almost *passé*.

Yet, despite the expansion and fragmentation of their subject, there has until now remained a common ethos amongst biochemists, a common method of approach to their problems that distinguishes them from their colleagues in other, adjacent disciplines. A biochemist (at least, the sort this book is about) is not just a physiologist applying chemical tools to living things, nor an organic or physical chemist who is interested in the properties of the chemical substances of the cell. Both such types of scientist exist, and the biochemist will most likely get along very well with them; but at the same time they are recognizably different, asking different questions and demanding different standards from the answers they get. Biochemistry has arisen at the meeting point of many sciences, and has been fertilized and enriched by all, yet it is itself essentially unique.

Faced with a living cell, or the tissue or organ which is composed of several million of such cells in close proximity to one another, the types of question the biochemist asks can be summarized under four major heads:

(1) What is the composition of the cell, in terms of individual chemical compounds which can be recognized as functionally different from one another and which can be separated by the techniques of chemical and physical fractionation?

(2) What are the relationships between these chemicals, and how are they made and converted one into another by the cell?

(3) How are these chemical interconversions controlled and regulated within the cell so as to enable it to maintain its organized structure and activities?

(4) What distinguishes the cell studied from those of other tissues, organs, or species? That is, in what way is its design related to the function that the cell performs within the living organism considered as a whole?

These questions, of course, all interconnect. Although they are arranged on the page in order of increasing complexity, it is not always necessary, or easier, to answer (1) before (2) or (3) before (4). An attempt to solve (2) may provide clues to (3) instead. But, nonetheless, they do represent separate passages in the biochemist's mind, and generally a different approach is required for each. In fact, a case may be made that they are also those questions which have been asked, broadly speaking, at different stages in the development of biochemistry as a science.

Thus, in the first phase of the history of biochemistry, the most important and pressing problem was the establishment of the nature and composition of the chemicals of the body. To this phase belong the work of Wöhler and Sumner already referred to, and the massive development of the French and German schools of organic chemistry in the hands of such nineteenth-century giants as Berthelot, Liebig, and Fischer, which resulted in the identification and subsequent synthesis of most of the simpler chemicals utilized by living organisms. The attack on the biological macromolecules (those with molecular weights of 10,000 upwards to a million or more), such as proteins, fats, carbohydrates, and nucleic acids, required a different order of techniques to those available a hundred years ago, despite the classical and painstaking analyses of such workers as Thudicum, who, in 1884, listed in his *Treatise on the chemical constitution of the brain* some 140 individual constituents, many of them complex combinations of fats, proteins, and carbohydrates, isolated by methods of extraction involving prolonged subjection of the tissue to conditions so extreme in acidity, alkalinity or temperature as to curl the hair of a contemporary biochemist with horror. Indeed a long battle was fought, lasting through until the 1920s, against those who believed that there *were* such giant molecules, rather than merely loose associations, *colloids*, of smaller units.

Today, the isolation and characterization of complex proteins has become a standardized operation. The first detailed molecular structure of a protein was worked out for the hormone insulin by Frederick Sanger in Cambridge in 1956, after nearly a decade of minute and laborious analysis which well deserved the Nobel Prize with which it was received; these procedures too have now become a routine piece of laboratory technique. Important though many remaining problems may be, the great days of the age of biochemical analysis are now truly past. In this book, we discuss its findings in Chapters 1 to 3.

Meanwhile, as the scale on which the biochemists could work moved up, that on which physiologists and microscopists were working moved down. The advent of the electron microscope for routine laboratory use in the 1950s united the two. It made possible magnifications of 100,000 times and more, and enabled the inside of the cell to be studied visually in great detail for the first time, so that biochemists could see in detail, as well as imagine, the inside of the cell whose properties they were investigating. As will be shown in Chapter 4, it became possible to deduce just *where* within the cell particular substances were located, and *where* particular reactions occurred. It became apparent that the cell was not merely a bag of randomly distributed chemicals, but that each substance had its own place and position. To many, the cell now seems to resemble more closely the regularly patterned form of a complex, hierarchically ordered set of macromolecules than the primitive sack of 'protoplasm' that had been the nineteenth-century picture.

The second phase of biochemistry was one of *kinetics*, of drawing a route-map setting out the major pathways by which chemical transformations occur within the living cell, and of understanding at a molecular level the mechanism of each individual chemical reaction. A primitive experiment of this sort was Spallanzani's, already described. The recognition in the early nineteenth century of the phenomenon of *catalysis*, in which chemical reactions are assisted and accelerated by substances (catalysts) which are themselves unaltered during the reaction, led to the assumption by the Swede Berzelius, in 1836, that the vast range of chemical activities occurring in living tissues depended upon the existence of potent chemical catalysts within the cell. And so indeed it proved. The heated controversies which followed, and which found distinguished chemists lined up in bitter opposition over whether the chemical reactions characteristic of life could be performed in the test-tube in the absence of living organisms, were resolved by the brothers Buchner, who, in 1897, ground yeast with sand in a mortar and extracted from the mixture a distinctly dead juice which was nonetheless able satisfactorily to ferment sugar to produce alcohol.

The name *enzyme* was coined for the catalyst in the yeast-juice that performed this desirable function, and the properties of enzyme soon showed that it was protein in nature. As more and more catalysts were discovered and extracted from the cell, the name enzyme became accepted as a general one for the entire class of biochemical catalysts. It is now recognized that practically every chemical reaction that occurs within the body requires its specific enzyme to catalyse it. Each enzyme catalyses only a single reaction, and the complete synthesis or degradation of a complex substance – for example, the breakdown of the starch in food to the sugar molecules of which it is composed during digestion in the gut, followed by the absorption of the sugar into the cells and its synthesis there to glycogen ('animal starch') or breakdown to carbon dioxide and water – requires a whole series of enzymes acting in sequence, one after the other. A reaction chain of this sort is called a metabolic pathway. The mapping of these pathways, for the synthesis and breakdown of sugars, fats, and amino acids (a few of which are outlined in Chapters 8 and 9), was the work of the generation of biochemists of the 1930s, and the names of Krebs, Embden, Meyerhof, Warburg, and Dickens stand high in this respect.

One problem remained in the understanding of these metabolic chemical interconversions: that of the *energy-balance* of the reactions. Destructive reactions (sometimes called '*catabolic*'), such as those of the breakdown of the sugar, glucose, to carbon dioxide and water, release considerable quantities of energy; synthetic (or '*anabolic*') reactions, such as the manufacture of proteins or fats, are energy-requiring. It is necessary for the cell to strike a balance between energy-producing and energy-demanding reactions, for it cannot afford to run for long at either a profit or a loss. It was F. Lipmann, of New York, who showed, in 1941, that the cell runs a sort of energy-bank, which can trap and store the energy released by catabolism, and provide it again on demand for anabolism, and that this bank consists of the chemical adenosine triphosphate (ATP for short). The significance and properties of ATP, as they are at present envisaged, will be discussed in detail in Chapters 5 and 7.

The 1950s saw a change in emphasis from the analysis of biochemistry-as-kinetics to that of biochemistry-as-information. The theoretical rationale for this transition was provided by the growth of the new sciences associated with the development of computers. Theories of 'control', 'feedback', and 'information transfer' were collated in 1948 by the American engineer and mathematician Norbert Wiener under the name of 'cybernetics'. As more and more became known about the mechanisms of individual enzymic reactions, about their energy-requirements, and about the workings of series of enzymes in the harmony of metabolic pathways, biochemists seized on these new concepts in order to probe the ways in which the cell controlled and regulated its own metabolism; how, so to speak, it decided at any one time

17

how much glucose to break down to carbon dioxide and water, or *how much* new protein to synthesize. And the triumph of bio-chemistry-as-information-flow was of course the spectacular solution to the problem of the mechanics of the accurate replica-tion of giant molecules such as DNA and of the translation of the genetic messages coded for in the DNA into the structure of the proteins themselves, undoubtedly one of the key scientific devel-opments of this century (Chapter 10).

This picture of the cell as a self-regulating mechanism, contin-ually changing, yet continually unchanged, is one of the most important and significant results of the new biochemistry of the 1950s and 1960s (Chapters 11 and 12). In the mid nineteenth cen-tury, the great French physiologist, Claude Bernard, had described the fundamental property of life as that of the ability to 'maintain the constancy of the internal environment'. Living organisms responded to exterior events impinging upon them in such a way as to absorb the effects of these events into their systems as rapidly, and with as little disturbance, as possible. They needed constantly to renew themselves, to recreate from within those portions of themselves destroyed in the rough-and-tumble of existence. The pattern of their bodies was fixed, although its individual components were forever changing. For this process, the name 'homeostasis' – 'staying the same' – was invented.

It has become the task of today's biochemistry to transfer this concept from the body as a whole to the working of each individual cell within it. Only when this has been done does it become pos-sible to define in chemical and physical terms how the organism as a whole, as the sum of its constituent cells, and all their myriad interactions, with each other and with the external world, can function. Armed with such knowledge, we can go on to ask the most fundamental of all questions: 'What is life, and how did it arise?' And to answer this is the prime aim of biochemistry.

CHAPTER 1

BEFORE WE START

In this book we shall be describing the biochemical make-up and behaviour of the living organism in some detail. In order to do this, we need constantly to use certain chemical words, phrases, and ideas. To those who have studied chemistry, what follows here is familiar territory; they would do better to skip the rest of this chapter. But, for those who are unacquainted with its jargon, there are included here a few brief paragraphs of definition in the hope that, having been disposed of, they need not trouble us unduly later on.

Chemists work with substances, which they attempt to purify one from another by making use of differences in their physical properties. For example, a mixture of salt and sand is separated because salt dissolves in water and sand does not; later, the salt can be recovered by boiling off the water. A substance which cannot be split by such physical methods into separate components is a *compound*. All chemical compounds, and there are many hundreds of millions of them, are formed by combination, in varying proportions of two or more, of a small number (about a hundred) of chemical *elements*; the elements can neither be converted into each other nor split into simpler substances by chemical means. The elements are represented by symbols, C for carbon, O for oxygen, Na for sodium, etc.; the compounds are indicated by a combination of these symbols. For example, common salt – sodium chloride – is NaCl.

Atoms and molecules

The smallest particle of an element is an atom, although the atom itself has an internal structure made up of smaller particles, protons, neutrons, and electrons. Neutrons and protons are packed together in the atomic nucleus. Neutrons are so named because they have no electrical charge, protons are positively charged, whilst electrons (sometimes written e^-) have an equal though opposite charge to the protons, making the atom neutral. The electrons are distributed between very precisely defined orbits or energy levels at varying distances away from the positive nucleus (they can be thought of like planetary orbits around the sun). Each of the energy levels are known as valency shells, and there is a maximum number of electrons that can exist in each shell. Atoms can combine with each other in fixed proportions to form molecules and when this happens they do so in a way that attains this maximum number in their outermost valency shell, as this gives the atom greatest stability. The number of electrons a particular atom has to lose or gain to achieve this stability is described as the valency of the atom. There are two ways this gaining or losing of electrons can be done: either by transferring electrons between atoms, or by sharing them. For example, the sodium atom, Na, achieves stability by *losing* an electron, thus becoming positively charged. The Cl atom does it by gaining an electron to become negatively charged. Occasionally, two or more atoms of a single element may join together *covalently* to form a molecule of that element; thus the gases hydrogen (H) and oxygen (O) normally exist as molecules containing each two atoms, H_2, O_2. The terms *atomic weight* and *molecular weight* are used to express the relative weights of atoms and molecules compared to the weight of hydrogen, which is the lightest element and whose weight is arbitrarily defined as 1. Thus the atomic weight of oxygen is 16, meaning that the oxygen atom is 16 times as heavy as the hydrogen atom. The molecular weight of water (H_2O) is $(2 \times 1 + 16)$ or 18. In general, the more complex a molecule, the larger its molecular weight. Salt (NaCl) has a molecular weight of 58, the sugar glucose ($C_6H_{12}O_6$) of 180, and some proteins of a million or more.

Ions, electrovalent bonds, and buffers

Charged atoms or molecules are called ions. If Na and Cl are close together in solution, transfer of electrons between them easily occurs, and because of their opposing charges, they are mutually attracted, combining to form salt, NaCl. The bond between them is described as electrovalent. In solution in water, electrovalent compounds such as NaCl tend to separate into their component ions, and to join together again only in the solid material, thus:

$$NaCl \rightleftarrows Na^+ + Cl^-$$

The opposite-headed arrows of this equation indicate that the reaction may, depending on circumstances, proceed in either direction. It is said to be *reversible*. When it is going from left to right, NaCl is said to *dissociate*; in the opposite direction, Na and Cl are referred to as *associating* to form NaCl. Other reactions – for example the *oxidation* (addition of oxygen) of coal gas (carbon monoxide) to form carbon dioxide, which occurs when we light a gas fire – are *irreversible* under normal circumstances:

$$2CO + O_2 \rightarrow 2CO_2$$

Compounds formed by the combination of two ions, such as NaCl, are called *salts*. Chemically, they are usually produced during the reaction of two other compounds, an *acid* and an *alkali*. Acids are substances containing the hydrogen ion, H^+ (for example hydrochloric acid, HCl). Alkalies, on the other hand, contain a negative ion, the *hydroxyl* ion OH^-, itself a combination between hydrogen and oxygen. An example is sodium hydroxide (caustic soda), NaOH. Acids and alkalies react in solution to give a salt plus water, thus:

$$NaOH + HCl \rightarrow NaCl + H_2O$$

A further word should be said here about the concept of acidity and alkalinity, because although a full treatment of it properly belongs to text-books of physical chemistry, biochemists find they cannot get very far without taking it into consideration. An acid solution, as we know, contains hydrogen ions, an alkaline one, hydroxyl ions. Both hydrogen and hydroxyl ions are generated by the ionization of water:

$$H_2O \rightleftharpoons H^+ + OH^-$$

This is a reversible reaction, and in reversible reactions the relative concentrations of the reactants always tend towards a constant equilibrium point. Therefore the amounts of hydrogen and of hydroxyl ions present in any aqueous solution must always bear a constant relation to one another. If we know the number of hydrogen ions present, the number of hydroxyl ions follows automatically. Thus an alternative way of regarding the *addition* of hydroxyl ions is to look at it as a *subtraction* of hydrogen ions from the solution. We can then define *both* acidity *and* alkalinity in terms of the concentration of hydrogen ions present in the solution; the more hydrogen ions, the more strongly acid the solution is. The hydrogen ion concentration of a solution is measured on a scale called the pH scale. The scale (which is logarithmic) runs from 0 to 14, 0 being the acid, and 14 the alkaline, end of the scale. Midway between the two, at a pH of 7·0, represents neutrality.

When hydrogen ions are added to a solution, the pH decreases; when they are removed (or hydroxyl ions added), the pH rises. But when certain substances are present in the solution, they will tend to combine with any H^+ or OH^- ions added, and, by removing them, they act so as to prevent the change in pH that would otherwise occur. Substances that act in this way are called *buffers*. In their ability to mop up hydrogen or hydroxyl ions, buffers act as regulators of pH. Such regulation is extremely important to the delicately balanced living cell, where sharp fluctuation in acidity and alkalinity can easily spell disaster.

Covalent compounds

But by far the largest number of substances with which we shall have to deal are not salts at all but compounds of the element carbon. Carbon compounds are so universally distributed amongst living organisms, and are so numerous, that their study has been split from those of other chemicals under the special name of *organic chemistry*. The carbon atom attains stability most easily by sharing its electrons with four atoms of a monovalent element or two atoms of a divalent atom. The simplest organic compound is thus methane (marsh gas) which has the formula CH_4. It is often useful when discussing carbon compounds, though, to abandon the simpler notation of inorganic chemistry and to try to draw a picture of the molecule as it actually exists in space. Thus is a two-dimensional drawing of the three-dimensional methane molecule, demonstrating that the carbon atom in methane is in fact entirely surrounded by hydrogen atoms. Really, though, even this is only an approximation to the actual three-dimensional structure of the molecule, which, if one had a microscope powerful enough to see it, would appear to be not flat all, but pyramid-shaped:

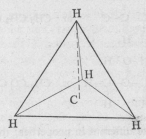

All the formulae we shall draw are, like that of methane, merely two-dimensional pictures of the real three-dimensional shape of the molecule.

In deference to the fact that the rule says that carbon must always be linked to four other atoms, four linking lines (*bonds*) are always drawn stretching out from any carbon atom. In cases such as that of carbon dioxide, CO_2, where the carbon atom is linked only to two atoms of oxygen, we draw the molecule as

$$O=C=O$$

showing that it is linked to oxygen not by *single* but by *double bonds*.

This type of bonding, by sharing electrons, is called *covalent*. Once made, covalent bonds are hard to break, and the substances which contain them therefore tend to be rather durable. In certain circumstances one partner atom of a covalent molecule may tend to obtain a greater or lesser share of the electrons than other partners. There is then a distribution of electric charge within the molecule which gives it a certain polarity, an important

The Chemistry of Life

property, as we shall see in relation to macromolecules (electrovalent molecules are of course completely polarized).

Drawing pictures of the molecules in space also reveals certain other characteristics about them. For instance, when we examine the substances corresponding to the formula C_3H_6O, we find that two possible structures exist, represented by

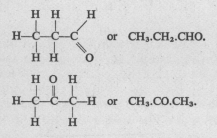

These two are quite different in properties: the first is *propionaldehyde*, formed by certain bacteria and a reactive, acrid-smelling substance; the second is the sweet-smelling *acetone*, familiar as the volatile base of nail varnish and aeroplane dope. The *structure* and *arrangement in space* of the molecules of organic compounds is thus critical to their behaviour. Substances (like proprionaldehyde and acetone) whose overall chemical composition is the same but whose structure and shape are different are called *isomers*. *Isomerism* is a common occurrence amongst the chemicals of the living body, and we shall meet it frequently.

Yet another property of carbon atoms is their ability not only to link together in long, straight chains such as those we have been drawing, but also to form *branched chains*:

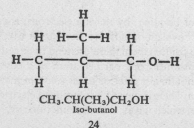

$CH_3.CH(CH_3)CH_2OH$
Iso-butanol

24

or even *rings*:

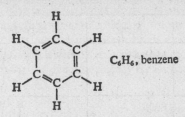

C_6H_6, benzene

Whenever we are dealing with molecules containing several carbon atoms, it is often useful to be able to refer to one in particular of the atoms. In order to do this, we *number* the carbon atoms in a molecule, starting at one end, preferably that nearest to some special group in the molecule, and going on to the other. Thus in *acetic acid*, we number the atoms starting with the acidic *carboxyl* group:

$$\underset{2}{\text{CH}_3}\underset{1}{\text{COOH}}$$

The advantage of such a numbering system will become obvious when we have to talk of larger, more complex molecules.

We shall discuss many more properties of carbon compounds during the course of this book, but however complicated they are, it is always the case that their structure and behaviour, from the simplest organic acid to the largest and most complex of proteins or nucleic acids, are governed by the same general rules that we have outlined here. There is nothing mysterious or inexplicable about the chemicals of the living organism. They may be very complicated and unstable and require special methods of handling, but the laws that they obey are those of chemistry and physics and no others.

With this preamble, we can move now straight to a discussion of the first of the sections into which we have divided biochemistry, that of the chemical nature of the substances that compose living cells, and which could be called *biochemistry as analysis* (Chapters 2 and 3).

TABLE 1. *Frequently occurring organic groupings*

Formula	Name	Expanded formula
.CH₃	Methyl	
.CH₂.CH₃	Ethyl	
.C:C.	Double bond	
.CHO	Aldehyde	
.CH₂OH	Alcohol	
.CO	Ketone	
.COOH	Acid	
.CH₂NH₂	Amine	
	Benzene ring	

CHAPTER 2

THE SMALL MOLECULES

We regard the human body as solid enough. In everyday life, we prefer to know about its interior workings only what can be deduced by looking at its outside. The smooth skin of the perfect body is a delight, and we are content if we can trace underneath it the muscles of our arms or the pattern of our ribs. It requires some conscious effort to recollect, at least while we are healthy, that we have lungs, pancreas, liver, and kidneys, and that, dissected out, the organs that compose our body, that are our person, would not look so dissimilar to those of the animals we eat. Yet despite our distaste at the process ('Last week,' as Dean Swift remarked, 'I saw a woman flayed, and you will hardly believe how much it altered her person for the worse'), it is not beyond our imaginative powers.

The effort required by biochemists is greater. For them the body is not composed even of organs, but of an astronomical number of individual cells, each only visible under the microscope. And each cell, in its turn, must be regarded as composed of many distinct chemicals, all capable of being purified, crystallized, of having their molecular weight and structure determined, and finally, ultimate indignity, of being stored in a bottle on a shelf along with innumerable others of far less sacred origin. When biochemists visualize the body thus, they at once become conscious that despite the seeming dissimilarity between, say, liver and brain, or a snail and a human being, the common likenesses between the cells that compose them and the chemicals that make up the cells far outweigh the minor differences.

Against this reduction of the body to chemical terms a long and bitter struggle was waged – yet it was a crucial stage in the development of scientific understanding of the nature of life, quite apart from its importance in the development of medicine. The alchemists had known that if one concentrated urine one could crystallize out urea, and that the distillation of ants in a retort produced the pungent formic acid. And it had been known since antiquity that the fermentation and distillation of vegetable matter provided alcohol. Such chemicals, and many others, had been known long before the phase of research that we have called 'biochemistry as analysis' began. And when during the nineteenth century chemists began to transmute into biochemists, handling with increasing skill the always more complex materials they extracted from living tissues, the list became ever longer.

To determine the proportions of the various chemical elements present in the human body is comparatively simple. All that is needed is an incinerator, a balance, and the variety of techniques that analytical chemistry has developed to deal with the simpler problems presented by minerals. For instance, a flame photometer can recognize the metallic elements by the difference in the colour of the flame produced when a small sample of each is burnt in air. There are also specific ion electrodes that can quantify the amount of particular elements present in a solution.

Table 2 shows the results of such an analysis. Three elements, oxygen, hydrogen, and carbon, are present in overwhelming preponderance. The last figures in the table show why. The vast bulk of body-weight is provided by water, thus accounting for most of the hydrogen and oxygen.

The many other chemicals of the body are nearly all compounds of carbon, also with hydrogen and oxygen, sometimes with nitrogen, occasionally with sulphur. Most of the calcium and phosphorus is combined as calcium phosphate and forms the hard substance known as bone. But for all this it remains true that, as J. B. S. Haldane said, even the Archbishop of Canterbury is sixty-five per cent water.

The analysis of just how the elements present are combined is a more complex task, and one that is in no sense complete even today. It is possible, though, to describe the *classes* of compound present with some precision, even though not all is known about each individual member of the classes. The crudest, and primary, division may be illustrated by an experiment in which a piece of tissue, such as liver or muscle, is ground in a pestle and mortar with cold dilute acid and the resulting purée (or homogenate, as it is officially, if inaccurately, described) filtered to separate the

TABLE 2. *The composition of the human body by weight*

Class	Substance	% body weight
As elements	Oxygen	65
	Carbon	18
	Hydrogen	10
	Nitrogen	3
	Calcium	2
	Phosphorus	1·1
	Potassium	0·35
	Sulphur	0·25
	Sodium	0·15
	Chlorine	0·15
	Magnesium, Iron, Manganese, Copper, Iodine, Cobalt, Zinc }	Traces
As water and solid matter	Water	60–80
	Total solid material	20–40
As types of molecule	Protein	15–20
	Lipid	3–20
	Carbohydrate	1–15
	Small organic molecules	0–1
	Inorganic molecules	1

soluble from the insoluble materials. The soluble fraction now contains nearly all the low-molecular weight substances present in the tissue (those, that is to say, with a molecular weight of anything up to 10,000 or so), whilst the insoluble residue, typically whitish or pale brown in colour and rather pasty in consistency, contains the high-molecular weight substances (with molecular weights of up to several millions), the proteins, nucleic acids and polysaccharides, as well as a variety of fatty substances.

For most biochemists, it is the giant molecules which are the really interesting and exciting ones, but even the most complex of giant molecules is built from, and its properties depend upon, a variety of smaller subunits, and it is these that we must begin by looking at.

LOW-MOLECULAR-WEIGHT COMPOUNDS

Soluble in the dilute acid of the experiment, and normally in solution also in the intact cell, are, first, a number of simple inorganic ions, notably the positively charged *potassium*, *sodium*, *calcium*, and *magnesium*, and the negatively charged *chloride* and *phosphate*. These ions are crucial for the functioning of the cell; they maintain its correct internal environment in the absence of which the macromolecules will be unable to function or may even disintegrate, and they are essential cofactors for many of the chemical reactions of the cell, as well as being directly involved in such cell functions as nerve transmission and muscular contraction.

Phosphates

The phosphate ion is an atom of phosphorus combined with four atoms of oxygen and has the formula PO_4^{3-}. It is significant because, unlike the other inorganic ions, it combines easily and enthusiastically with many organic compounds, due to its strong electronegativity. Because of its negativity it attracts protons (H^+ ions) very easily and in fact usually exists within the cell as orthophosphoric acid H_3PO_4:

but we shall frequently shorten the formula, and write it simply as Ⓟ. It can readily form phosphate *salts* with, for example, sodium or potassium, or phosphate *esters* with organic alcohol. *Monoesters*, *diesters*, and *triesters* are theoretically possible:

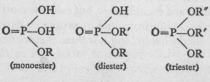

30

The Small Molecules

(where R stands for any organic alcohol) but in our extract we are likely to find only the monoesters and diesters. All the organic compounds which contain the alcoholic group —CH_2OH can form such esters, and we shall come across *sugar* phosphates, *amino acid* phosphates, *hydroxyacid* phosphates, *amide* phosphates and *nucleoside* phosphates. The phosphate ester has the power of making an otherwise relatively inert organic compound biochemically very reactive indeed, and we shall find time and again, when the cell synthesizes or transforms an organic molecule, that the first step in the process is to convert it into its phosphate ester.

Phosphate groups can also combine, to give *di-* and even *triphosphates*, thus:

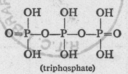

(triphosphate)

Such di- and triphosphate esters are also common biologically, and are even more reactive than the monophosphate esters.

Organic acids

The simple carbon compounds listed below are acids, although weak ones, because they readily ionize in H_2O to give protons and a negatively charged group. The simplest is acetic acid and these acids can be purified from extracts of most living tissues. Acetic acid is of course in vinegar, and citric acid is found in citrus fruits and sherbet. They are chemically speaking not very exciting substances. Yet in life they are of great significance because of their ability to take part in complex chemical reactions with ease and speed. Starved of them, and of the sugars, the wonderfully complex machinery of protein and lipid that composes the cell comes tumbling down.

Fatty acids

These acids are vitally important to the cell as they fulfil an important structural role combined as lipids in the cell membrane.

TABLE 3. *Some commonly occurring acids*

Name	Type	Formula
Acetic	Monobasic	$CH_3.COOH$
Succinic	Dibasic	$CH_2.COOH$ | $CH_2.COOH$
Fumaric	Dibasic	$CH.COOH$ || $CH.COOH$
Lactic	Hydroxyacid	$CH_3.CHOH.COOH$
Malic	Hydroxyacid	$COOH$ | $CHOH$ | $CH_2.COOH$
Citric	Hydroxyacid	$COOH$ | CH_2 | $HO\!-\!C\!-\!COOH$ | CH_2 | $COOH$
Pyruvic	Keto acid	$CH_3.CO.COOH$
Oxaloacetic	Keto acid	$CO.COOH$ | $CH_2.COOH$

The fatty acids are comparatively simple molecules containing unbranched hydrocarbon chains about 14–24 C atoms long together with an acidic COOH group. The predominance of the long hydrocarbon over the relatively small acidic group means that the fatty acid is essentially non-polar; that is, because of the symmetrical distribution of the electron shells there are no electrical charges. The fatty acids do not dissolve easily in water – they are *hydrophobic*. The hydrophobic nature of fatty acids has important implications for their role in membrane structure and the ways in which they interact. Another important property of the fatty acids is their degree of saturation; that is the number of double C bonds in their chain structure. Whilst single bonds in a molecule do not impose any constraint on its shape, because the parts of the molecule are free to rotate around them, double bonds are fixed and cannot be rotated. Thus two molecular configurations are possible for each double bond:

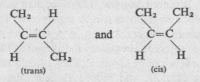

(trans)   (cis)

While in the trans position the symmetry of the groupings keeps the molecule straight, in the cis position the chain becomes bent and changes direction. Straight chain saturated fatty acids, and trans unsaturated acids, can pack together tightly, whilst kinked unsaturated ones cannot stack so neatly, and thus lipids formed from such acids are more fluid.

The acidic, polar group of the fatty acids is also important as it can combine easily with other organic molecules, as we shall see when we consider lipids later.

Sugars

Sugars can be written according to the general formula $C_nH_{2n}O_n$; in the simplest case, $n = 3$, and we can write two possible isomeric formulae:

CH$_2$OH.CHOH.CHO   glyceraldehyde (an aldose)
CH$_2$OH.CO.CH$_2$OH   dihydroxyacetone (a ketose)

Sugars which contain the aldehyde group —CHO are *aldoses*, those which contain the ketone group —CO are *ketoses*. If the formula for glyceraldehyde is drawn more fully, it can be seen, however, that it hides a further complexity. *Two* possible structures for glyceraldehyde can be drawn, which differ from one another in no other respect but that one is the mirror image of the other. These structures are *isomers* (see page 24):

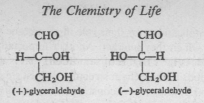

(+)-glyceraldehyde   (−)-glyceraldehyde

That these two are not identical will become clear if one imagines trying to rotate one of the drawings so as to superpose it directly on the other. It cannot be done, any more than mirror-images or right- and left-handed gloves can be superposed. They thus form a special class of isomers, known as *optical isomers*. Although these two molecules are chemically indistinguishable, their physical behaviour is not quite identical. They often form asymmetric, mirror-image crystals, and indeed Pasteur, the discoverer of optical isomerism, first separated the two isomeric forms by picking out the different shaped crystals. Also, and most important, they differ in the way they interact with plane-polarized light.*

We can draw the molecules of higher sugars, where n is greater than 3, by adding successive CHOH groups between carbon atoms 2 and 3 of glyceraldehyde or dihydroxyacetone. When this is done, it is found that each new carbon atom creates a new centre of asymmetry in the molecule, and can hence become the focus for further isomers. By the time $n = 6$, when the sugars are collectively known as *hexoses*, there are sixteen possible isomers for the aldoses and another sixteen for the ketoses. Fortunately, the cell is very discriminating. Half of each set of isomers, those based on (−)-glyceraldehyde, never appear at all in nature, whilst of the others only a very few are at all common. Amongst the aldoses, the most important are glucose, galactose, and mannose. Amongst the ketoses we need only mention fructose which, though based on (+)-glyceraldehyde, rotates plane-polarized light to the left.

* In particular, when plane-polarized light is passed through a solution of one of the two isomers of glyceraldehyde, it is found that the plane of polarization of the light is rotated; each of the two mirror-images, however, rotates the plane in opposite directions. Viewed through one isomer the plane of the light is rotated to the right (+), whilst through the other it is rotated to the left (−). This power of rotating the plane of polarization of light is common to all molecules which are, like glyceraldehyde, asymmetric – that is, can be drawn as non-identical mirror-images.

6 CHO	6 CHO	6 CHO	6 CH_2OH
5 |	5 |	5 |	5 |
HCOH	HOCH	HCOH	C=O
4 |	4 |	4 |	4 |
HOCH	HOCH	HOCH	HOCH
3 |	3 |	3 |	3 |
HOCH	HCOH	HCOH	HCOH
2 |	2 |	2 |	2 |
HCOH	HCOH	HCOH	HCOH
1 |	1 |	1 |	1 |
H_2COH	H_2COH	H_2COH	H_2COH
(+)-galactose	(+)-mannose	(+)-glucose	(+)-fructose

The hexoses have the important property of being able to form ring-type structures; drawn diagrammatically, the process of ring-formation can be shown like this:

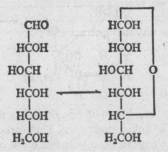

(the two molecules have identical formulae). But a better way of showing the ring is like this:

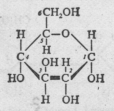

We shall in future frequently represent the glucose molecule this way. Dissolved in water, nearly all the glucose is in this closed-ring form (99·976 per cent to be precise). Drawn like this, the nature of the asymmetry of the CHOH groups can be seen more clearly; some of the —OH groups stick up above the plane of the ring, whilst the rest fall below the plane. All the other hexose isomers have a slightly different pattern of 'above-and-belowness' for the —OH groups.

This however creates another asymmetric C atom at C1. The OH group on C1 can be either *below* or *above* the plane of the ring, in the α or β position. This type of stereoisomerism is important because it is C1 that is the starting point for the polymerization of sugars into polysaccharides. Thus the particular configuration at this atom can affect the whole shape of the polysaccharide chain and hence its biological function.

A similar type of ring-closure occurs with the ketose sugars such as fructose, but in this case the ring is a five-membered one, thus:

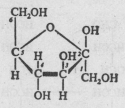

The sugars can combine with one another in their ring forms to produce *chains* of linked sugar units. Thus carbon 1 of one molecule links, via an oxygen atom, with one of the other carbons of a second molecule, generally carbon 4. Such a bond is known as a *glycosidic linkage*. An example is the compound between one molecule of glucose and one of fructose:

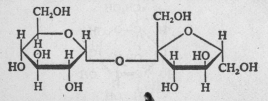

Here the bond is between C1 and C4, and in the α position, so the molecule is 1–4 α, and the substance is better known as sucrose – common table sugar. Similarly, two glucose molecules may combine to form *maltose*:

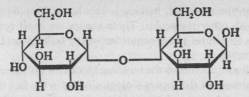

This process of linking (polymerization) can be repeated indefinitely, to build up chains of sugar units. Such chains, which may be several hundreds of units long, are *carbohydrates* (polysaccharides), the first of the giant molecules we shall shortly discuss.

Although the most important sugars are those containing 6 carbon atoms ($n = 6$), there are some examples of 7-carbon sugars, and rather more of 4- and 5-carbon sugars. When $n = 5$, the sugars are called pentoses, the most important of them being *ribose*. Like the hexoses, the pentoses normally exist in ring form and ribose can be drawn like this:

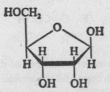

2-deoxyribose, in which the carbon atom at position two has lost its oxygen, also occurs. We shall shortly meet both ribose and deoxyribose as important constituents of the complex *nucleotide* molecules.

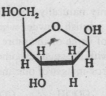

More complicated sugars exist because of the capacity for substituting ring hydroxyls with various other groups, e.g. NH_2, to give galactosamine, or $NHCOCH_3$ to give N-acetylgalactosamine. These substitutions can considerably change the electric charge on the molecule, making it capable of forming stronger bonds with other molecules. These sugars are most commonly found in structural polysaccharides where stability and greater rigidity matter.

As well as their role as the building blocks for polysaccharides, the importance of the sugars biologically lies in the fact that they are the normal source of most of the energy utilized by the body; it is by the oxidation of glucose that the cell obtains the energy it needs for all the rest of its activities.

Amino acids

The amino acids are the building blocks of which the giant protein molecules are composed. Each amino acid contains nitrogen as well as carbon, hydrogen, and oxygen, and their general formula can be written:

where R may be any one of a number of different groups; in the simplest case, R is hydrogen and the amino acid is glycine:

There are about twenty naturally occurring amino acids in all; each, when pure, is a whitish powder with a faint but distinctive smell. The formulae of some of the more important of them are listed in Table 4. As with the sugars, isomeric forms of the amino acids can exist, as mirror-images of one another, thus:

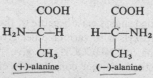

Also, as with the sugars, only one form is the naturally occurring one; in this case the one based on (−)-glyceraldehyde. The other isomers are neither produced nor utilized by animal cells, though they are used by some bacteria.

Inspection of the formula of the amino acids shows that they contain both an acidic group (the carboxyl, COOH) and an alkaline group (the amino-group NH_2). In a neutral solution both of these groups are ionized, and it would be more correct to write the formula of an amino acid in water as:

Depending on the nature of R and the pH of the solution, amino acids may be positively or negatively charged, or neutral. The presence of different R groups can also considerably affect the electrical charge on the molecule, making the amino acid predominantly electropositive or electronegative, e.g. the carboxyl side groups of aspartate and glutamate make them negative ions whilst lysine and arginine have positively charged NH_3^+ groups. The ionic properties of the amino acids mean that they can act as buffers (p. 22) in solution.

Purines and pyrimidines

Just as amino acids are the starting points for the proteins, so another group of giant molecules, the nucleic acids, find their building blocks in substances derived from two simple related ring compounds, *purine* and *pyrimidine*.

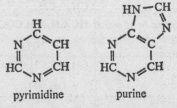

pyrimidine   purine

The most frequently occurring derivatives of pyrimidine are *cytosine*, *uracil*, and *thymine*, whilst the commonest purines are

The Chemistry of Life

TABLE 4. *Some common amino acids found in proteins*

Name	Abbreviation	Formula
Alanine	Ala.	$CH_3.CH.COOH$ $\quad\quad\|$ $\quad\quad NH_2$
Aspartic acid	Asp.	$HOOC.CH_2.CH.COOH$ $\quad\quad\quad\quad\|$ $\quad\quad\quad\quad NH_2$
Cysteine	Cyst.	$HS.CH_2.CH.COOH$ $\quad\quad\quad\|$ $\quad\quad\quad NH_2$
Glutamic acid	Glut.	$HOOC.CH_2.CH_2.CH.COOH$ $\quad\quad\quad\quad\quad\|$ $\quad\quad\quad\quad\quad NH_2$
Glutamine	Glun.	$H_2N.O.C.CH_2.CH_2.CH.COOH$ $\quad\quad\quad\quad\quad\quad\|$ $\quad\quad\quad\quad\quad\quad NH_2$
Glycine	Gly.	$CH_2.COOH$ $\|$ NH_2
Leucine	Leu.	CH_3 $\quad\searrow$ $\quad\quad CH.CH_2.CH.COOH$ $CH_3\nearrow\quad\quad\quad\|$ $\quad\quad\quad\quad\quad NH_2$
Phenylalanine	Phen.	⬡$—CH_2.CH.COOH$ $\quad\quad\quad\quad\|$ $\quad\quad\quad\quad NH_2$
Proline	Prol.	$\quad NH$ $\quad\diagup\diagdown$ $CH_2\quad CH.COOH$ $\|$ $CH_2—CH_2$
Serine	Ser.	$HO.CH_2.CH.COOH$ $\quad\quad\quad\|$ $\quad\quad\quad NH_2$
Threonine	Thre.	$CH_3CH.CH.COOH$ $\quad\quad\|\quad\|$ $\quad\quad OH\ NH_2$
Tyrosine	Tyr.	$HO—$⬡$—CH_2.CH.COOH$ $\quad\quad\quad\quad\quad\|$ $\quad\quad\quad\quad\quad NH_2$

adenine and *guanine*. The bases differ from each other because of the different groups substituted into the rings. It is these substitutions, together with the shape and spatial orientation of each of the rings, that make possible the specificity of base pairing and determine the physical characteristics of the nucleic acids, as we shall see in the next chapter.

Mostly the purines and pyrimidines do not occur free, but are combined with the sugar, *ribose*. Generally, one or more phosphate groups are also present. Such molecules, containing purines or pyrimidines, ribose and phosphate, are known as *nucleotides*, thus emphasizing the part they play in the formation of the nucleic acids (see page 67).

Depending on the number of phosphate groups present (it will be remembered that we have already shown how phosphate groups

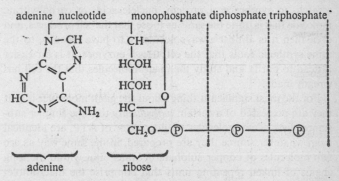

can readily be converted into polyphosphate groups), nucleotide mono-, di-, and triphosphates are possible. As was mentioned previously, the addition of phosphate groups to a substance has the effect of conferring biological reactivity on it, and, in conformity with this rule, the nucleotides, and especially their triphosphates, are amongst the most highly reactive of any the biochemist has to deal with, and they are utilized by the cell in many important reactions, especially biosynthetic mechanisms. Of all the chemicals to which we shall refer, none will be more frequently mentioned than adenine nucleotide triphosphate, which we shall henceforward abbreviate to ATP.

CHAPTER 3

MACROMOLECULES

The low-molecular-weight compounds that we have described occupy an intermediate zone between organic chemistry and bio-chemistry. Given time and the right optical isomers to start with, the organic chemist can synthesize most of them by classical tech-niques, and the physical chemist can determine their structures precisely; both can be reasonably certain that the molecule with which they are working is in all respects identical to the one found in the living organism. But with the giant molecules, the position is not so simple; it is difficult, though not impossible, starting from scratch, to synthesize a protein, nucleic acid, or carbohydrate molecule that is quite the same as those that the cell produces, and even then it is almost always necessary to have recourse to the same synthetic tools that the cell uses – enzymes. It is far easier instead to purify and study the macromolecules that the cell has already made.

For the most significant thing about the giant molecules is that they are possessed of a certain *individuality* that the simpler sub-stances lack. Two molecules of glucose, or of ATP, are identical from whatever source they are prepared, in the same way as are two molecules of copper sulphate or water. But within the long chains of linked repeating units that comprise the macromole-cules, there is room for variety and permutations of great subtlety, enough to puzzle chemists and their classical armoury of analytical tools. Indeed it is only in comparatively recent years that these giant molecules have been studied in their own right, as the earlier biochemists believed them to be simply aggregates of the smaller molecules that we have already discussed.

However, what is confusing to the analytical scientist is much less so for the cell, whose precision in distinguishing tiny varia-tions in molecules whose weight may be in the order of millions is as great as the analytical power of the best of computers. The classical case is that of haemoglobin, a blood protein which con-tains some three hundred amino acids; changing the sequence of the chain, by swapping just one of these amino acids for another, is sufficient to prevent the haemoglobin from functioning properly, and results in the disease known as sickle-cell anaemia. We shall meet many similar cases, for such specificity of structure is one of the major differences between living and non-living things.

The giant molecules are the stuff of life; our muscles, skin and hair are protein fibres; beneath our skin, giant molecules of lipid are laid down in layers of subcutaneous fat; around us, the bark of trees and the stems of plants are long, close-packed molecules of the carbohydrate cellulose. The feel of a living thing is the feel of a complex meshwork of interacting macromolecules.

The chemist puts them firmly in their place, defining them briefly as substances with molecular weights of 10,000 and upwards, with somewhat indeterminate physical properties, composed by the polymerization (joining together in long chains) of simple low-molecular-weight units. But beyond saying this, chemists tend to regard them as rather outside their domain. So we have come at last to the realm of biochemistry proper.

Biochemical methods of dealing with the macromolecules tend to be somewhat *ad hoc*; they have rather to take them as they find them. So we will increasingly find that we have to abandon something of the classical outlook of the chemist. It is difficult to ask of a macromolecule 'is it a pure substance?' for one has few criteria of purity. One extracts the molecule from the living tissue, and in doing so one is forced to disrupt the delicately balanced unity of the cell, to submit it to acids, salt solutions, organic solvents, ion-binding resins, and many other such chemical probes. When we finally have left a substance which we cannot break down any further without doing something *really* drastic, like hydrolysing* or oxidizing it, and which we cannot resolve by such physical techniques as the application of high gravitational or electrical fields, we are at liberty to announce a 'pure' protein, lipid, or whatever. But we may be sure that this purity is one that we have imposed by our own operational demands on the living material, and that in the cell the interconnections between the macromole-

---

* 'Hydrolysis' means the splitting of a molecule by adding the elements of water, H_2O, to it. It is generally carried out by heating the substance in solution with dilute acid or alkali.

cules we have arduously separated are quite as important as the composition of our pure sample. But, especially in the first stages of biochemistry, there was no choice. Before asking *how* the cell functions, we must first know of *what* it is made.

And so we purify the macromolecules, analyse them by breaking them down into their constituent parts, study their structure, and plot the sequence in which the units are joined together. The job of biochemistry as analysis stops there. When, finally, we can put the molecule back into the cell from which it came and try to describe how and why it behaves as it does in any particular biological situation, then we have entered another phase of research.

PURIFICATION AND STRUCTURE
OF MACROMOLECULES

Methods of separation may exploit the differential solubility of the macromolecules in different solutions, for instance, salt solutions of different strengths, or in organic solvents (alcohol or acetone); the different stabilities of proteins to acids and alkalies; the different mobilities of the proteins under gravitational and electrical fields (centrifugation and electrophoresis); and the ability of certain substances to absorb some proteins but not others (chromatography). Some of these methods are worth describing in a little detail.

Certain artificial resins, and some materials such as cellulose and aluminium hydroxide, tend, when stirred with a solution containing charged ions, to draw the ions out of solution and bind them to themselves. If the resin is packed into a glass cylinder (column) and the solution poured in on top of the resin column, the liquid that emerges at the bottom is practically free of solute. Such a principle lies behind several domestic water-softening devices. If a protein solution is used, the proteins become bound on the column. They can, however, be washed out again by increasing quantities of water or salt solution. This is the principle of *chromatography*. Each protein passing down the column runs through at a characteristic speed, depending on the concentration and pH of the solution being used to wash it out. Thus the different proteins in the original mixture arrive at the bottom of the column one after the other, and, if the liquid emerging from the column is collected in separate tubes, a few drops at a time, some tubes will contain one of the proteins, some another, and a separation will have been obtained. This elegant method was first used by the Russian botanist Tswett for separating the different coloured materials present in plant pigments. Modern developments of chromatography make possible the separation of proteins based not only on the ionic charge on the molecule but also upon its shape and size.

Gel filtration enables one to separate the molecules from one another and at the same time gives an estimate of their molecular weight. It involves washing a solution of the macromolecules through a glass column packed with a gel-like substance composed of small beads of porous material. Large molecules pass through the column easily because they cannot enter into the beads. Smaller molecules, however, can penetrate the gel pores, and therefore take a more circuitous and longer route through the column. If a solution of a substance of known molecular weight is also passed through the column, an estimate of the molecular weight of the separated macromolecules can be obtained by comparing the rate at which the substances pass through and are washed out. Molecular weight can also be accurately determined by the process known as gel *electrophoresis*, if it is carried out in the presence of a detergent, S.D.S. (sodium dodecyl sulphate). The detergent binds to the macromolecules, making them negatively charged. If an electric current is applied to a gel containing a mixture of such macromolecules they will migrate towards the positive electrode. Their rate of progress will depend on size and can be compared to the electrophoretic mobility of a substance of known molecular weight.

The next step after separating a macromolecule and obtaining a 'pure' sample is to try to determine its shape, an important factor in any consideration of the biological function of the macromolecule. Especially with polysaccharides this can be done by investigating their behaviour in solution. Solutions of different shaped molecules differ in properties such as viscosity, which depend upon the particular shape of the molecule concerned.

X-ray diffraction is by far the most accurate of the methods for determining shape, having the ability to discriminate between individual atoms in a large molecule. A sample of the solid substance is bombarded with X-rays which will be scattered as they hit the atoms. If these diffracted rays are directed on to a photographic plate, a pattern can be developed which corresponds directly to the orientation and distribution of atoms within the molecule. We shall see this technique 'in action' later in this chapter.

The final piece of information needed to determine structure can only be obtained by breaking the molecule down into its constituent parts to enable its components to be identified and to plot their sequence and the nature of their linkages.

Certain properties, common to all types of macromolecule, can be derived from such analyses. All seem to have a certain individuality and many have the ability to specifically recognize and interact with other substances. They are all, to a greater or lesser extent, flexible molecules, capable of adapting their shape in response to their particular microenvironment within the cell. They all have a pronounced intolerance to extreme conditions, a fragility that makes them fall apart if treated harshly. Finally they all may be described in terms of a structural hierarchy of primary, secondary and tertiary structures that we need to look at in a little more detail.

The primary structure of a macromolecule refers to the large number of similar building blocks, each covalently bonded together, giving a stable, firm backbone to the whole molecule. Secondary structure describes how this backbone is folded or pleated in a regular way to allow a large number of weak internal bonds to be formed between different chains or different parts of the same chain. As a general rule, secondary structure is determined by a macromolecular chain folding in such a way as to maximize the number of *hydrogen bonds* it can make. (This type of bonding takes place whenever two electronegative atoms, usually O and N, come close enough together to interact with hydrogen and share a positive charge.) There are a variety of other weak bonds which help shape the macromolecular structure, but it is not necessary to discuss them all in detail here.

Tertiary structure describes the way in which these H-bonded chains twist in on themselves to give a complex 3-D shape. A variety of weak bonds and ionic interactions between different parts of the long molecular chain hold it in a tight and well-defined structure, with parts of the molecule buried deep inside, inaccessible to water or other small molecules, while other regions form the surface, where they may interact with all the other ions and molecules which swarm in the microenvironment of the macromolecular chain. Now although the combined strength of many of these weak bonds is quite great, each individual bond requires little energy to be broken or remade and it is this fact that gives the molecules their flexibility. Breaking a few weak bonds will slightly change the shape of the molecule, without significantly altering its overall properties. The importance of this will become clear later. This kind of conformational change may also provide the cell with a method for getting rid of unwanted molecules by exposing previously protected inner areas to degradative attack.

Paradoxically, although these weak bonds are a source of strength and are an essential aspect of the molecule's structure, they are also responsible for the fragility of the molecules, because the bonds depend for their existence on very precise conditions of ionic strength, pH, temperature and the extent to which water is present within their localized area in the cell. A change in any one of these factors may cause destruction of the bonds and an unfolding of the molecule into a naked single chain. This process is called denaturation and is generally irreversible.

With these general comments, we can now turn to a consideration of the particular classes of macromolecules in more detail.

CARBOHYDRATES (POLYSACCHARIDES)

We saw earlier that the hexoses, glucose and fructose, could combine by means of a glycosidic linkage to form molecules each containing two sugar units, *disaccharides*. The glycosidic link is formed through the carbon atom 1 of one sugar and the carbon atoms 2, 3, 4, or 6 of a second. The formation of a disaccharide leaves free carbon atom 1 of the second sugar, thus:

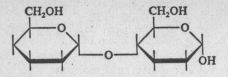

and a further glycosidic link becomes possible:

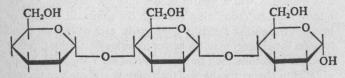

This process can be repeated indefinitely. Three sugar units in a row make a *trisaccharide*; when there are more than three the polymer is called an *oligosaccharide*, and when there are many (many being defined in this case as more than ten) it is a *polysaccharide*. As in forming a glycosidic linkage, each sugar molecule loses a molecule of water, polysaccharides have the general formula $(C_6H_{10}O_5)_n$. For a typical polysaccharide such as starch, n may run into the hundreds; in cellulose it can be anything from 300 to 2,500.

The polysaccharides are widely distributed in plants and animals, both as structural substances (e.g. cellulose), and as food storage compounds (starch, glycogen). As the number of units in the chain increases, the polysaccharide changes its physical properties. Mono- and disaccharides such as glucose and sucrose are water-soluble and taste sweet; as the molecular weight increases, solubility decreases and the taste alters; a 'starchy' taste is quite different from sugar. After a certain point, however, it no longer seems to make much difference how many units more are in the chain. The addition or subtraction of ten, twenty, or even a hundred seems scarcely to alter the properties of the molecule. Nor does it ever become possible to refer to 'the' molecular weight of a polysaccharide; a handful of starch will be made up of many chains of varying lengths. The best that can be given is an average figure, and even this tends to differ depending on the source of the sample and the method of purification.

It is clear that, within the polysaccharide chain, there is considerable scope for variation. Thus, as well as the number of units within the chain, it is possible also to alter

(1) any or all of the individual sugar units (galactose for glucose, for example);

(2) any or all of the points of attachment between the sugars (carbon 1 to carbon 6 instead of to carbon 4, for example);

(3) the geometry of these attachment points (for example, whether the linkage is in the α or β configuration, as discussed previously – see page 36);

(4) by linking carbon 1 of one sugar to one of the carbon atoms of a glucose which is already part of a chain, it is possible to produce *branched chains*. Such branches can be one unit or several units long, and the branches can themselves divide further, like those of a tree (Figure 1a).

Depending on how these links are made, the resultant macromolecule will differ considerably in properties. Obviously, the nature of the sugar units (and whether they have amino groups attached to them) will affect the chemical behaviour of the macromolecule. But in addition whether the glycosidic bond is α or β, for example, will affect the flexibility of the molecule and whether individual molecules can lie close together in a rigid configuration.

So how does one set about discovering the structure of a molecule in which so many permutations are possible? The temptation to despair when faced with a molecule which can be composed of 1,000 units, any one of which may be one of sixteen different hexose isomers (to say nothing of the pentoses) linked in any one of five possible ways, each with an isomeric twin, to its neighbour, must be resisted. Despite the fact that the number of possible permutations that such variations provide is rather greater than those one is faced with on the average pools coupon, only a relatively small number actually seem to occur, and, even amongst these few naturally occurring combinations, minor variations do not seem to be very significant for the state of the final product. But one may well blench at the problem of actually determining the structure of these mammoth molecules. Fortunately, though, as we have seen, quite elegant methods do exist.

The first step is obviously to determine which of the monosaccharide sugars are actually present, and this can be done easily enough by hydrolysing the polysaccharide (quite strong hydro-chloric or sulphuric acid is often necessary). On hydrolysis, all the glycosidic bonds between the sugar units break, releasing the constituent sugars, which can then be separated and identified. If only one sugar is present (when the polysaccharide is called a *homo*-polysaccharide) then one is nearly home and dry. If two or more sugars are present, it becomes necessary to determine the details of the order in which they are joined together. Fortunately, it is usually the case that the various sugars are linked to form a regular, simple, repeating sequence; for example, an alternating series of glucose and galactose units. Sequences are studied by a modification of the hydrolysis procedure, in which milder conditions and shorter hydrolysis times are used. Under such circumstances the polysaccharide chain is not entirely broken down to monosaccharides, but is instead only partially hydrolysed into many small fragments of chain, each containing several sugar units. Just as with the total hydrolysate, the different fragments of the broken chain can be separated and the structure of each determined.

To take a hypothetical case, suppose the original polysaccharide contained glucose and galactose in equal proportions. We might then find amongst the chain fragments the oligosaccharides

    gluc-gluc,
    gluc-gluc-gluc,
    gluc-gluc-gluc-gluc, and,
    gluc-gluc-gluc-gluc-gal,

and also,

    gal-gal,
    gal-gal-gal, and,
    gluc-gal-gal.

We might then suspect that the original polysaccharide was composed of separate glucose and galactose chains, with a gluc-gal branching point where one chain joined the other. Alternatively, if the products of hydrolysis were the fragments

    gluc-gal-gluc,
    gal-gluc-gal,
    gal-gluc-gal-gluc, and so forth,

one would clearly be dealing with a polysaccharide in which each chain contained an alternating sequence of glucose and galactose residues.

In more sophisticated versions of this 'jigsaw' technique, the biochemist seeks for specific enzymes which will hydrolyse bonds between some sugar residues but not others. In all such methods, however, one ultimately has to solve the intricate puzzle of fitting the bits together again in what seems the most likely manner – a task which, these days, computers make rather easier.

To complete our picture of the polysaccharide, a figure for the molecular weight is needed. Some information is provided by the chemical evidence, if we know chain lengths and numbers of branching points, but physical methods of the sort we have discussed in the previous section are normally indispensable.

The use of such techniques has provided the basis for our understanding of the composition of polysaccharides. New and hitherto unknown polysaccharides are constantly being found as more and more plant, bacterial, and animal tissues are analysed, and many of these have highly complex patterns of constituent sugars and branching chains. But the most frequently occurring of the polysaccharides are rather simpler in structure. There are three: two of plant origin, *cellulose* and *starch*; and one from animals, *glycogen*.

Cellulose

Think of a naturally occurring fibre, and it is odds-on that it will be made of cellulose. Cotton, flax, wood, paper, even rayon are all cellulose; the hairs of the cotton seed plant, in particular, are over ninety per cent pure cellulose. There is more of it in the world than any other organic chemical, for it forms the structural framework on which plants (and some bacteria) are constructed.

Hydrolysis of cellulose yields nothing but glucose. Partial hydrolysis gives glucose 1–4 β-linked di- and trisaccharides. So cellulose has a primary structure composed of straight chains of 1–4 β-linked glucose units; molecular weight estimations, which give a value of 50,000 to 400,000, imply that there are between 300 and 2,500 glucose residues per molecule.

The value of cellulose as a natural fibre lies in just this fact of

The Chemistry of Life

long, straight glucose chains. It is the β-linkage that gives the cellulose chain its ability to lie like a flat ribbon, enabling molecules to be packed together, all lying in the same direction, to form a crystal-like thread with a strength somewhat greater than a thread of high-grade steel of the same diameter. The long threads are composed of oriented bundles of cellulose molecules, each individual bundle being known as a *micelle*. There are also H-bonds formed between the glucose units of the individual molecules within the micelle, thus increasing its rigidity. Individual micelles can easily be seen when electron microscope pictures of cellulose fibres are made, and crystallographers, by measuring the dimensions of the micelles, have been able to deduce a great deal about the intra-molecular configuration of cellulose fibres. From the point of view of the biochemist, interested in natural structures, it is the rigidity of the fibres which is of most interest, because it is this that enables cellulose to provide the framework for the plant cell. When the uniform orientation of the molecules is lost, and the micelles are shuffled so as to point in many different directions, the fibre loses its strength; the product of this disorientation has commercial value as *cellophane* (Figure 1c).

Apart from cellulose, many other polysaccharides help maintain plant structure. Among these are *arabans* and *xylans*, which give wood many of its typical properties and are made from the pentose sugars arabinose and xylose.

Chitin serves the same function in an insect exoskeleton as cellulose does in the plant. Chitin differs only in that one of the ring hydroxyls in each of the repeating glucose units is replaced by the acetamido $NHCOCH_2$ group, but this small difference means that more H and other bonds can now hold the chains together, making for even greater rigidity. The acetamido group also bonds very easily to proteins and in fact chitin is usually found linked to protein, to form a proteoglycan.

Starch

Starch typifies the role of the polysaccharides as food stores in plants; storage organs such as tubers, fruits, and seeds may contain up to seventy per cent of starch, which can be broken down into glucose and used for food when the need arises. It is easy to see

Macromolecule

FIGURE 1. *Carbohydrate molecules*

(a) Sugar units can be arranged in several ways:

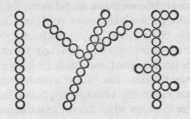

(b) In cellulose the long, straight molecules lie side by side:

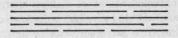

(c) While in cellophane they are disorganized:

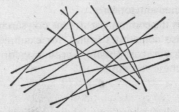

(d) In amylopectin the molecule is highly branched:

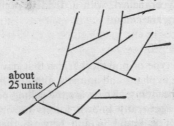

about
25 units

how this storage role is aided by the use of a molecule which is compact, insoluble but readily degradable. Unlike cellulose, starch is a granular material with no trace of organized crystalline structure, yet it too is composed of chains of glucose units, 1–4 linked. The critical difference lies in the fact that the chains in starch are 1–4 α-linked. Starch is a mixture of two polysaccharides, the one a straight chain called amylose, about 250–300 glucose units long, with the chain folded and packed down into a helical shape. The other is a branched chain called amylopectin in which straight 1–4 linked chains are joined together by 1–6 linked branching-points, one every twenty-five glucose units or so, to produce a molecule which when drawn looks more like a gorse bush than anything else (Figure 1d). It is these branching points which make the molecule easy to degrade, as they provide sites for the beginning of breakdown, as we shall see later. About seventy to eighty per cent of starch, by weight, is made up of the branched chain amylopectin. The two chains can be separated simply by adding hot organic solvent to a starch dispersion, when the amylose is precipitated. The ability of starch to form pastes depends on the highly branched amylopectin.

While starch is the commonest storage polysaccharide, *fructose* polymers are also sometimes found, for example in Jerusalem artichokes and asparagus, possibly accounting for the rather distinctive taste of these vegetables.

Glycogen

Animals store polysaccharide mainly in the liver and muscle, as *glycogen*. This is a glucose polymer similar to amylopectin, but even more highly branched, with a dividing point every 12–18 glucose units along the chain.

Other polysaccharides

There is another type of polysaccharides that have a supportive or protective role in plants and animals, although by a totally different method to cellulose. These are the gelling polysaccharides of which pectin (generally known for its use in jam-making) is an example. It is found in and between the cell walls of higher plants and is composed of chains of galactose, arabinose, and the sugar acid galacturonic acid. The glucosaminoglycans are found in the mucous linings of the respiratory and digestive tracts of animals. Hyaluronic acid and chondroitin sulphate provide the lubrication in knee joints and the agars prevent seaweeds drying out at low tide. These gelling polysaccharides differ from the fibrous molecules in that the covalently bonded portions that make up their primary structures are connected by areas where there is a three-dimensional network of weakly bonded units, often forming helical shapes. This 3-D network can trap large quantities of water, forming a gel, and this confers elasticity on the material, enabling it to stretch and then bounce back to its original shape. This is of obvious importance where a protective coating must not restrict movement, such as inside the stomach and in the cell wall of red blood cells, where it enables them to squeeze through the narrow capillaries.

Polysaccharides do have important roles at the surfaces of animal cells, where their structural differences have been exploited during evolution in the development of complex phenomena whereby cells when they come into contact 'recognize' one another so as to distinguish like from unlike – an important property in relationship to the body's immune system. However, in general they play a less important structural role in animals than in plants. This role is filled in animals by lipids and proteins, which offer greater potentialities for physical and chemical modification than the carbohydrate polymers, and hence enable the animal to achieve a more flexible and adaptable existence. It is perhaps significant that the only animals to use polysaccharides as structural components are those, like insects and arthropods (crabs, for instance), where the tough, brittle chitin is used as a so-called external skeleton. It is just this existence of the exoskeleton which has placed such sharp limitations on insect size. For animals at least, there is no future to being built on a polysaccharide framework.

PROTEINS

In 1838, the name protein was given by Mulder, at the suggestion of the chemist Berzelius, to the group of nitrogen-containing materials that he had found in animal and plant materials. The

55

The Chemistry of Life

name comes from the Greek for 'first things', and it was a forunate choice. Most of life revolves around the activities of the proteins; the energy of the cell goes largely first into making them, and then into using them to perform a multitude of diverse jobs. So much so, indeed, that a measure of the importance of the proteins is that we should be hard put to it to decide whether they evolved to suit the demands of living, or whether life developed to fit the requirements of the proteins.

What are they? Macromolecules which on hydrolysis yield practically nothing except a mixture of amino acids. In the protein, the amino acids are linked by *peptide bonds*. The peptide bond is formed by joining the amino-group of one amino acid to the carboxyl of a second. Here is the equation of peptide bond formation between glycine and alanine.

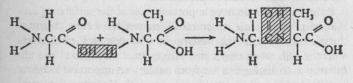

glycine   alanine   glycylalanine+water

The peptide bond, then, is the linkage

$$-\overset{\displaystyle |}{\underset{\displaystyle \parallel}{C}}-\overset{\displaystyle |}{\underset{\displaystyle |}{N}}-$$

between two amino acids, and the resulting molecule is called a *peptide*. The linkage is covalent, and the peptide bond fills the same function for proteins as does the glycosidic bond for polysaccharides.

As two amino acids are present in the molecule shown in the equation, it is specified more precisely as a *di*peptide, in this case glycylalanyl dipeptide. The peptide link between the two amino acids still leaves free the —NH₂ amino-group of the glycine, and the —COOH carboxyl group of the alanine. Both of these are thus

56

free to participate in further peptide bond formation with other amino acids. For example, with serine, either of the two reactions

glycylalanine + serine → glycylalanylserine + H_2O (2)
serine + glycylalanine → serylglycylalanine + H_2O (3)

is possible. The two tripeptides, though both containing the same amino acids, are different compounds and will have different physical and chemical properties. For a tripeptide, then, six different isomers are possible, the two shown, and also alanyl-serylglycine, alanylglycylserine, serylalanylglycine, and glycyl-serylalanine. (We can in future abbreviate these cumbersome names by using the standard diminutives shown in the list of amino acids, Table 4, page 40). From these formulae, it is apparent that each peptide contains an N-terminal end and a C-terminal end, and the convention is to write the formulae starting with the N-terminal.

Such peptides can be prepared by rather tedious organic chemical synthetic routes – early in the century Emil Fischer scored a notable triumph by making an octadecapeptide with eighteen residues, only to be subsequently outstripped by Abderhalden and Foder who produced one with nineteen. But such achievements, the products of many years of work by some of the most brilliant organic chemists who ever lived, fade into insignificance when one faces the naturally occurring polypeptide chains, each containing several hundred amino acids. Such polypeptides are commonly known as proteins; the distinction between a long polypeptide and a short protein is purely verbal. Peptides with only a few amino acids (generally two or three) do occur in the cell, and there are some peptide hormones with eight to ten. But the biochemist's major concern is with the proteins.

We saw that for a tripeptide, six isomeric structures were possible. As the number of amino acids increases, the number of possible isomers grows enormously. Synge has calculated that for a relatively modest protein – with a molecular weight of 34,000, and with 288 amino acids, but made up of only 12 different amino acids out of the possible 20 – the number of different isomers is 10^{300}. If only one molecule of each isomer were to exist, the total mass would be some 10^{280} grams. As the weight of the earth is only 10^{27} grams, it is very clear that only a tiny fraction of these isomers in fact exist. Even so, the number of proteins in existence is a formidable one. Several hundreds have been more or less purified, and it is entirely likely that the total number of chemically and physically distinguishable proteins in any one living species runs into the tens of thousands, whilst there is also a good deal of evidence to show that even apparently similar proteins from different species are not quite identical in structure and amino acid composition. So that, within the compass of life on earth, there must be several millions of individual and unique proteins.

But here we run into the prime difficulty of protein biochemistry – deciding precisely what we mean by an 'individual and unique protein structure'. Granted that every polypeptide chain with a defined amino acid sequence is a single protein, and that any chain which differs from it by so much as a single amino acid altered or out of place is a different one (and we mentioned earlier the evidence of sickle-cell anaemia to show that this is in fact the case), how do we distinguish between the chains, how do we obtain a pure protein?

The analytical techniques that we have discussed, of gel filtration, chromatography and electrophoresis, are all important here, where it is vital to use methods which do not destroy the delicate protein chains in the process of handling them.

These tools are continually being superseded by new advances; new analytical techniques are constantly revealing that once 'pure' proteins are in fact composed of several different components, and the definition of purity alters decade by decade and almost year by year.

Nowadays, the battery of available techniques makes the purification of most soluble proteins a tolerably straightforward affair. The first necessity is to have some means of recognizing the specific protein one is trying to purify. This is easy if the protein is an enzyme, because then it, and no other protein, will catalyse a specific reaction, and the presence of the enzyme in a mixture of other proteins can thus always be noticed. Any protein fractions which do not contain the ability to catalyse the reaction do not contain the enzyme. Similarly the protein may be tagged if it is one of those containing heavy-metal ions, such as iron, which are firmly bound to it (*metalloproteins*); other such tagging groups are the phosphates of *phosphoproteins* and the carbohydrates of *mucoproteins*.

Having obtained the protein, how do we know if it is pure? A lot depends on exactly what we mean by pure in this context. In the early days of protein chemistry a lot of attention was paid to ridding the protein of all traces of non-protein materials, such as metal ions or fragments of lipid or carbohydrate. But it is now clear that in the living cell itself much extraneous material occurs bound into extremely close association with the protein. To remove it, even when this can be done, may produce a satisfactorily chemically pure amino acid chain, but that is not really what one is after; the important thing is to isolate as close an approximation to the native material as possible – even if this does mean taking it with its warts and all.

One can only tell if the protein is really free of other proteins by trying to go on fractionating it. If it doesn't split into several components under ultracentrifugation or electrophoresis or chromatography, it is probably a single molecular species. Pure proteins can normally be crystallized, but even this is not an absolute indication of purity; there may still be five or ten per cent of another protein present that cannot be removed without great difficulty. But only when one is sure that the protein is pure can any attempt be made to determine its structure.

As with polysaccharides, there are two parts to the structural problem, determination of the primary sequence, the order of amino acids along the protein chain, and the shape of the molecule in space, its secondary and tertiary structures.

Obtaining this information is not easy, and for only relatively few protein molecules are primary and secondary and tertiary structures all known. But at least it is now known *how* to solve the problems. The only saving grace in the matter of amino acid sequence is that polypeptide chains are never branched as polysaccharides are; there is only a linear sequence to cope with. This isn't to say that many proteins do not consist of several polypeptide chains; they do, but, as we shall later see, they are joined by linkages other than peptide bonds; they belong to the secondary structure of the protein.

The first step towards the amino acid sequence is to find out just *which* amino acids are present, and this is done by hydrolysing the protein and separating and determining each of the amino acids in turn. The quantitative determination of the up to twenty different amino acids present in a protein used to be a mammoth task lasting many months; over the last decade, however, the entire process has been automated, thanks to the advent of chromatography. Automatic amino acid analysers are now standard laboratory equipment. The only trouble is that the information provided by the amino acid analysis is of strictly limited value. By correlating it with the molecular weight of the protein, it is possible to decide just how many times each amino acid appears within the original protein molecule, but few deductions about the way they are linked can be made.

One must, then, resort to the 'jigsaw' techniques we mentioned in connection with polysaccharide structure, and try to build a picture of the complete protein by combining the information gained from an analysis of many small sections of it. It is of importance here to find which amino acids form the N- and C-terminal ends of the molecule. Fortunately, there exist certain enzymes and chemical reagents which combine specifically with *either* the C- *or* the N-terminal amino acid. Use of these enables one to decide which amino acid begins, and which ends, the protein chain.

Once the terminal amino acids are known, the next step is to make a series of partial hydrolysates, splitting the protein into fragments two to five amino acids long, and determining the sequences in each of these chainlets separately. Partial hydrolysates can be made with acid hydrolysis for short periods of time or at low temperatures, but there also exist several enzymes which break down protein chains, and these enzymes have the virtue of being fairly discriminating about which bonds they will and won't attack. Trypsin, for example, an enzyme present in digestive juices, splits any peptide bond where the amino acid at the C side of the CONH bond is either lysine or arginine, thus producing peptides whose C-terminal amino acid is lysine or arginine. Chymotrypsin, another digestive enzyme, does the same where the amino acids are phenylalanine or tyrosine. Many other protein splitting enzymes can be used similarly.

All that remains, when one has obtained the partial hydrolysates from these sources, is to separate them and determine each individual sequence before trying to put them all together in the most likely structure. Frederick Sanger and his associates at Cambridge completed this little puzzle for the small protein insulin, which has less than sixty amino acids and a molecular weight of only 12,000, in rather less than a decade, and in doing so became in 1956 the first people ever to publish the full sequence for any protein. By the time Sanger was given a Nobel Prize for this achievement in 1959, several other protein structures were far advanced, and complete or nearly complete sequences are now known for 500 or so proteins in many different species, the largest being more than 500 amino acid units in length. Meanwhile, the way towards a quite different approach to protein sequencing has been opened by increasing knowledge of the structure of DNA and RNA. We will see later how it is that DNA and RNA sequences determine proteins; for now the point is simply that one can predict a protein structure on the basis of a nucleic acid structure. In 1978 the first virus nucleic acid sequence was completely decoded, and hence it became possible to know the viral protein sequences too. This technique will be of increasing importance in the future.

How about the higher order structure of proteins? We have said that secondary structure is determined by simply folding the primary chain so as to achieve the largest number of H bonds between different parts of the chain. However, in the proteins another much more powerful bond is provided by the amino acid cysteine:

$$HS.CH_2.CH.COOH$$
$$|$$
$$NH_2$$

The —SH group in one molecule of cysteine can readily combine with that of a second to produce a 'disulphide bridge':

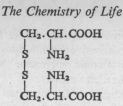

If the two cysteines are in different parts of a protein chain, or are in two separate chains, the chains bend towards each other

FIGURE 2. *Two protein chains*

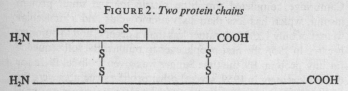

(a) Insulin: two chains bound together by disulphide bridges

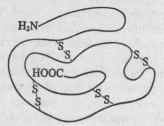

(b) Ribonuclease: one chain twisted into a ball by disulphide bridges

and are held together by the bridge. Many proteins contain such cysteine bridges which are covalent bonds and cannot easily be broken. Thus in insulin there are two disulphide bridges between the two chains of which the molecule is composed, whilst a third bridge connects two distant portions of one of the chains. In ribonuclease, which is a single-chain molecule, four disulphide bridges twist the chain into a snake-like configuration (see Figure 2), whilst the enzyme chymotrypsinogen has five bridges. Apart from disulphide bridges the other weak bonds we have mentioned are also present helping to maintain secondary and tertiary structure.

According to their shape, the protein molecules are divided into *fibrous* and *globular* proteins. As their names imply, the fibrous proteins have straight chains, the globular proteins have chains which are coiled together into irregular ball-shaped molecules. Amongst the fibrous group are the structural proteins of the body; keratin in hair and nails and wool, collagen in skin and connective tissue, elastin in tendons and arteries, fibroin in silk. These proteins are all insoluble, relatively tough molecules resistant to acids, alkalies, and quite high temperatures. The resistivity of these molecules, like that of cellulose, is a reflection of their ability to form ordered, semi-crystalline structures. The advantage to the biochemist is that it is possible to make X-ray photographs of these crystal-like formations and use them to deduce the shape of the molecules.

The simplest of the fibrous proteins to be studied was fibroin, from silk, which showed under X-rays a regular repeating pattern within the molecule, each unit of the pattern being 7 Ångström units (Å) long (an Ångström unit is 10^{-8} centimetres). As each amino acid residue is 3·5 Å long, this pattern was interpreted as being 2 amino acids in length, suggesting that the silk molecule could be drawn as a flat, zigzag chain. Each complete zigzag would be 2 amino acids long. Two zigzag chains, lying side by side, would stick together at alternate 'zags' by reason of hydrogen bonds.

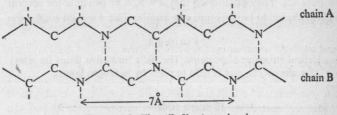

FIGURE 3. *The silk fibroin molecule*

Such a repeating pattern could result in an extended meshwork of chains, and a description of this sort is sufficient to explain the X-ray diffraction patterns made by fibroin.

Other molecules, though, are more complex. The classical experiments were made with keratin. The keratin fibre can, when

The Chemistry of Life

soaked in water, stretch to as much as twice its original length.
The unstretched form is called α-, the stretched form, β-keratin.
Astbury, at Leeds, who made these studies, found that the stretched
β-keratin gave an X-ray picture practically identical to that of silk
fibroin. The α-keratin, then, must be present in some more contracted form than that of the silk fibroin. Great efforts were made
to interpret the structure, and many ingenious suggestions put
forward, until, finally, in 1951, Pauling and Corey in America
concluded that the only pattern which met all the requirements
was if the molecule was coiled, like a loose spring, into a spiral
conformation which they termed the α-helix. The helix can be
shown diagrammatically as in Figure 4(a) or in more detail as in
Figure 4(b). In its natural shape the α-helix makes five complete
turns for every 18 amino acids, or one turn for 3·7 acids, and the
amino acids which fall immediately above and below each other
as the helix ascends clutch hold of one another by means of hydrogen bonds, and so keep the whole structure stable. When the
α-keratin molecule is stretched, the helix is straightened out like a
pulled spring, until ultimately it assumes the 'unwound' fibroin-
configuration of β-keratin. Pauling and Corey later refined their
picture of the helical protein molecule by showing that the helix
itself was not formed on a straight but on a helical axis, so that it,
too, gradually twisted, like the coiled-coil of a lamp filament
(Figure 4c). They postulated that it would be possible for several
helical chains to twist round one another like a woven rope – the

FIGURE 4.

(a) Helical structure of proteins. The helix turns five times for every
18 amino acids, and is held in place by hydrogen bonds.

(a)

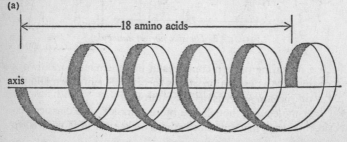

64

FIGURE 4 (cont.)

(b) The Pauling α-helix: a close-up view.

(c) Some super-helices.

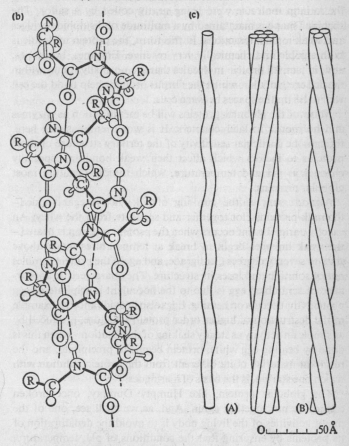

(b)   (c)

(A)   (B)

0 ⌞⌞⌞⌞⌞⌞ 50Å

protein in hair and nails, they suggested, was formed when six keratin α-helices wrapped themselves round a seventh into a pleated bundle.

All the fibrous proteins seem to be built on a pattern similar to that of keratin or fibroin. The helical configuration of α-keratin is also found in the globular proteins, but here the entire helix is bent and bunched together into a compact globule as if the taut rope of the keratin molecule were being neatly coiled by a sailor. The globular shape is maintained by a multitude of disulphide bridges and weak ionic interactions. In this form, the protein molecule is both soluble and chemically very reactive. Enzymes, hormones, and, in fact, all protein molecules that are normally found within the cell are globular, whilst the fibrous proteins help build the cell wall or fill up the spaces between cells.

Many of the globular proteins will be met in action as enzymes and hormones in later chapters. It is worth emphasizing here, though, the particular sensitivity of the tertiary structure of these proteins to factors which affect their weak bonding, especially changes in pH and temperature, which rapidly denature most globular proteins.

A good example is the 'curdling' of milk when vinegar is added – the milk proteins clot together and separate from the whey. An exactly parallel event occurs when the protein solution is heated – the weak linkages begin to break at temperatures much above sixty to seventy degrees Centigrade, and again the protein tumbles out of solution and loses its structure. The characteristic appearance of scrambled egg is due to the flocculent precipitate of egg proteins that occurs on heating. Eggs also provide another example of the destruction of higher order protein structure, produced by as simple an activity as steady shaking of the solution – when this is done by beating egg white, protein begins to precipitate, and the resultant material, quite different from the protein solution with which one started, is the basis of meringues.

The globular protein, like Humpty Dumpty, once broken cannot be put together again. And, as we shall see, one of the major activities of the living body is to avoid the denaturation of its proteins by ensuring that the conditions of pH, temperature, and so forth within the cell never exceed certain prescribed limits. In the absence of life, as that visionary Friedrich Engels pointed out a century ago, the proteins are inherently unstable.

Macromolecules

NUCLEIC ACIDS

The next class of giant molecules we have to consider is that of the nucleic acids. In 1868 Friedrich Miescher in Tübingen isolated from the nuclei of pus cells from discarded surgical bandages an unusual phosphorus-containing compound which he called *nuclein*. He went on to show that a similar substance was present in many other less distasteful materials, notably salmon sperm, and it became apparent that nucleic acids, as the new substances became known, were in fact widely distributed in living tissues. Although they were later found to be not entirely confined to the cell nucleus, the name remained as a general one for this entire group. When, much later, the role of the nucleic acids in heredity and in protein synthesis came to be understood, the name acquired, indeed, a new significance and justification.

One of the best sources of nucleic acids in animals was found to be the thymus gland, an organ which is of importance in young animals (especially calves), where it secretes a not-very-well-understood hormone, but which atrophies after the animal passes puberty. Nucleic acid from thymus gland was found to yield on hydrolysis a mixture of purine and pyrimidine bases, a sugar, and phosphoric acid. The purine bases were shown to be adenine and guanine, the pyrimidine bases cytosine and thymine, and the sugar deoxyribose – all substances that have been met in the previous chapter. Nucleic acid from bacteria, plants, and yeast was found to differ from thymus nucleic acid in containing uracil instead of thymine and ribose instead of deoxyribose. At first, it was thought that these differences indicated that there were two nucleic acids, one for animals, the other for plants, but it subsequently became clear that the ribose- and uracil-containing nucleic acid was present in animals as well, and, in fact, that all cells contain both types of nucleic acids. It is the deoxyribose-containing nucleic acid which is found in the nucleus of the cell; the other is more widely distributed. The two classes are now known as ribonucleic acid (RNA) and deoxyribonucleic acid (DNA).

Both DNA and RNA occur in the tissue in combination with varying amounts of protein, and such protein, especially that found attached to DNA in the cell nucleus, has quite characteristic properties; notably, it is very alkaline, containing large amounts of the amino acid arginine. Such proteins are known as histones, whilst the nucleic acid with protein attached is described as a nucleoprotein. In extracting the nucleic acid from the tissue, one of the problems is to remove the tightly bound protein without using methods so drastic as to harm the nucleic acids themselves. This problem is made easier by the fact that the nucleic acids are considerably more robust than proteins; they are stable to mild acids and alkalies, and to heating to almost one hundred degrees, procedures which rapidly wreck the delicate protein molecules. Thus RNA may be extracted from yeast by dissolving in a detergent solution at ninety degrees, whilst DNA can be prepared from thymus by extraction with alkali or salt solution followed by chloroform.

The material that is obtained by these methods is a whitish, fibrous substance (DNA rather resembles asbestos in appearance), quite stable, which can be stored for longish periods without coming to much harm. Its molecules have a weight in the order of tens of millions and, like proteins or polysaccharides, have a primary structure consisting of a repeating series of simple units joined into a chain. Although, as we shall show, many different DNAs and RNAs exist, the problem of isolating one DNA from a mixture of others, so critical in protein biochemistry, scarcely exists. This is at least partly because the physical properties of the different molecules do not differ sharply enough from one another to make it possible to exploit differences in solubility or stability as can be done with the proteins. Chromatography of DNA does reveal that the 'total' DNA obtained from any tissue can be split into a number of slightly differing fractions, but it has become customary to refer to the 'total' DNA of any organ almost as if it were a single molecular species – so one talks of thymus DNA, liver DNA, and so forth. In fact the properties of calf 'thymus DNA' do differ slightly from those of calf 'liver DNA', and those of rat kidney RNA from guinea-pig kidney RNA.

Protein consists only of amino acids, polysaccharides of sugars, but the nucleic acids contain purines, pyrimidines, sugars, and phosphoric acid. Thus the basic pattern of the repeating unit is less simple than that of the other macromolecules. It was early apparent that the base, sugar, and phosphate were combined as nucleotides, where the purine or pyrimidine is linked through the sugar to the phosphate (see page 41). How were the nucleotides connected to each other, though? The early hypothesis was that the four bases, sugar and phosphate groups were linked together in a so-called 'tetranucleotide' repeating unit, but such a concept was found to conflict with the experimentally observed base ratios in DNA and RNA.

Some relationship between the four bases must exist, though, because it was noticed by Erwin Chargaff in New York that for RNA the amount of (adenine plus cytosine) always equals (guanine plus uracil), whilst for DNA an even more precise relationship holds, so that the amount of purines and pyrimidines present in any sample is always identical and the ratio of adenine to thymine and of guanine to cytosine present is exactly one. Also, DNAs tend to fall into one of two main groups – either one in which there is more adenine and thymine than guanine and cytosine, or the rarer one in which guanine and cytosine are the two main components. So it is clear that the nucleic acids are highly organized and well-ordered compounds, even if the pattern of arrangement is not immediately obvious.

The layout of bases, sugars, and phosphates within the molecules was discovered during the 1930s and 1940s. The proof was a difficult one, involving some complex feats of organic chemistry, but the conclusions are now well established. For RNA, each nucleotide can be drawn like this:

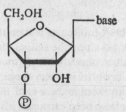

where carbon atom 1′ of the sugar is attached to the base, and carbon atom 3′ to the phosphate. When the nucleotides join to form RNA, the link is between the phosphate of one nucleotide and carbon atom 5′ of the second sugar:

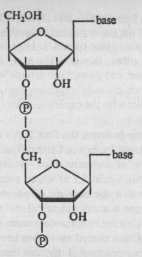

Thus the primary chain link runs through the sugar phosphates, and can be shown schematically:

sugar-base
/
phosphate
/
sugar-base
/
phosphate
/
sugar-base
/
phosphate

The variable in RNA molecules is then the order in which the four bases are arranged along the chain. Although there are only four bases, the number of available permutations down a chain of several hundred is obviously very large, though smaller than for proteins. What knowledge there is of the actual arrangement found in RNAs that have been extracted from the tissues depends on the type of analyses we are now quite familiar with. Hydrolysis, which splits the RNA into its residual nucleotides, gives a modicum of information; partial hydrolysis, in this case using a variety of enzymes, including ribonuclease and also some potent nucleic acid splitting enzymes from snake venom, is of more value. Ribonuclease from pancreas has an interesting specificity, for it seems preferentially to attack bonds between pyrimidine nucleotides and other bases. For example, the enzyme will split the nucleic acid chain

pupupy | py | py | pupy | pupupy | py |

(writing pu for purines and py for pyrimidines), along the lines drawn, to give 2 pupupy + pupy + 3py.

But despite the many ingenious methods that have been devised, and the enormous interest that there now is in the whole problem of RNA and DNA in relation to their role in genetics and in protein synthesis, only a few molecules have yet had their sequence reliably and completely determined, although in the next few years progress is likely to prove a good bit more rapid. The situation is better for a group of relatively low-molecular-weight RNA molecules which occur in solution in the cell, the so-called transfer RNAs (see Chapter 10) with about 80 nucleotides each, whose structure was worked out at all levels by American researchers.

The structure of the repeating unit of DNA is similar to that of RNA, with the simple replacement of ribose by deoxyribose. Thus, like RNA, the DNA chain is linked through a 3'—5' sugar phosphate bond, with the purine and pyrimidine bases tacked on to carbon atom 1' of each deoxyribose residue (see diagram on page 70 and Figure 5). But why, if this is all, the peculiar regular ratios that seem to exist between the amounts of the different bases? And what are the aspects of the secondary and tertiary structure of DNA, whose fibrous character and behaviour in solution are those of a rigid, linear molecule, like a stiff rod?

X-ray studies made by Rosalind Franklin and by Maurice Wilkins in London in the early 1950s indicated that there was a regular repeating sequence within the molecule every twenty-eight Ångström units. Now the distance between two nucleotide residues, from phosphate to phosphate, is only seven Å, so in some way there must be a strong repeating tertiary structure to the DNA. James Watson and Francis Crick, in Cambridge, solved the problem in 1953 (Wilkins, Watson and Crick were given the 1962 Nobel Prize for doing so, and Watson has written an entertaining, if disingenuous account of the affair in his book *The Double Helix*).

They found the answer in a helix – at almost the same time as Pauling and Corey found that the α-helix also gave the answer to the keratin problem for proteins. The DNA helix as proposed by Watson and Crick turns once every ten nucleotides (proteins, it may be remembered, manage to make the turn in 3·7 amino acids), and it contains a *double strand* of two DNA chains twisted together about the same axis. All the way along, the two chains link together *through their bases*. Now when Watson and Crick came to make scale models of the structure they proposed, they found that the two chains could *only* fit together if every adenine on one chain was opposite a thymine on the other, and every guanine opposite a cytosine. Then, and only then, the two strands would latch together, and hydrogen bonds between each pair of bases would hold them firmly in place. By contrast, if one considers adenine and cytosine, which do not form base pairs, the three atoms necessary for H bond formation cannot be set into a straight line, and hence the bond cannot form. The secondary structure of DNA is very robust because of the additive effect of many H bonds between the base pairs, all orientated in the same way, i.e. stacked on top of each other. Figure 5a shows base pairing between adenine and thymine, and how it works for one portion of the double chain is shown diagrammatically in Figure 5b. The helical tertiary structure is made possible by the orientation in space of the flat aromatic rings of the bases. If the double chain twists into a helix a large number of weak bonds can be formed between each ring and others lying directly above or below it. Again the sheer number of these bonds greatly enhances stability of the molecule, but they are nevertheless individual, weak and easily broken – we shall see in later chapters how very important this is to nucleic acid function.

In three dimensions, Watson and Crick drew their model as in Figure 5(c). The whole concept brilliantly accounted for the X-ray analysis and also for the strange constancy of the ratios of the amounts of the different bases that had been so puzzling. In the structure of DNA shown here, the amount of adenine inevitably equals that of thymine, and guanine that of cytosine, just as the chemical analysis had shown.

Despite the restrictions placed on the base-order in such a

Macromolecules

FIGURE 5.

(a) Mortice-and-tenon arrangement of base-pairs.

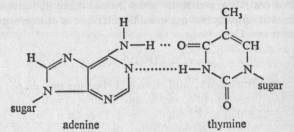

adenine   thymine

(b) Bonding of two chains in DNA helix.

sugar—A ... T—sugar
  | |
phosphate   phosphate
  | |
sugar—T ... A—sugar
  | |
phosphate   phosphate
  | |
sugar—G ... C—sugar
  | |
phosphate   phosphate
  | |
sugar—C ... G—sugar
  | |

(c) Watson-Crick DNA helix.

model, the number of permissible isomers is, in fact, the same as that of RNA – that is, astronomically large. Finding the order of the bases presents similar problems to those faced with RNA, and, like that molecule, they have not yet been solved. Perhaps the most hopeful way forward will lie in the fact that, as we shall see when we come to discuss the mechanism of protein synthesis, given the structure of a protein, it is possible to predict the structure of the DNA and RNA which helped make it. But such a discussion properly belongs to a different phase of biochemistry.

LIPIDS

We have left the lipids until last mainly because they do not conform to our concept of a macromolecule quite as well as the other three classes that have been discussed. They are a diverse group, with few points in common other than the property of being insoluble in water but soluble in organic liquids – a very operational definition! They are much smaller than the other macromolecules, having molecular weights in the hundreds instead of thousands or millions, and their inherent structural hierarchy is a little more difficult to define. This is because they are rarely found as individual molecules, because they have a tendency towards self-aggregation, and it is the aggregate that we must consider as the macromolecule, the building blocks of the primary structure being the individual lipid molecules themselves. However, the building blocks in lipids are not held together by strong covalent bonds but by weak bonding due to interaction between their dominant non-polar regions. The absence of a strong backbone and the presence of only weak bonding in the structure of lipid aggregates means that the macromolecular property of flexibility is much more strongly emphasized, so that it can almost be described as fluidity. Lipids, as a group, include, as well as fats, *esters*, formed by the combination of an alcohol with an organic acid. The ester linkage is produced according to the equation

$$R.COOH + R'.OH \rightleftarrows R.COOR' + H_2O$$

where R and R' may represent any organic radical.

Fats and Oils

The prototype fat is a combination of a straight chain fatty acid, sixteen or eighteen carbon atoms long, with *glycerol*, a sweet, sticky liquid containing three alcoholic groups which can combine with one, two, or three fatty acid molecules, yielding a mono-, di-, or tri-glyceride.

An *oil* differs from a fat only in being liquid at normal temperatures – that is, an oil is a fat with a low melting point.

The typical fat therefore has the formula:

$$CH_2OOCR$$
$$|$$
$$CHOOCR'$$
$$|$$
$$CH_2 . OOCR''$$

R, R', and R'' need not necessarily be the same, and this general formula can cover a variety of possible permutations. But in practice eighty per cent or more of fats contain one or more of the three acids:

palmitic $[CH_3(CH_2)_{14}COOH]$,
stearic $[CH_3(CH_2)_{16}COOH]$,
and oleic $[CH_3(CH_2)_7CH=CH(CH_2)_7COOH]$.

These three are simple straight chain acids; oleic differs from the other two in having a double bond (see Table 1, page 26). The fats are named according to the acids present, so that tristearin has three stearic acids, whilst if one oleic and two palmitics are found the fat is oleodipalmitin.

In a naturally occurring fat a mixture of many of the possible combinations will be found, although the exact proportions of each will depend on the source of the fat, and are not purely random; some combinations are preferred to others. In plants, double-bonded acids such as oleic tend to predominate – olive oil is almost pure triolein (plants also have a higher proportion of the more unusual acids than do animals). In animals, on the other hand, the single-bonded acids are more frequent.

As the melting point of the fat is directly related to the number of double bonds present in its constituent acids (the more double bonds, the lower the melting point) plant fats tend to be liquid at normal temperatures and are therefore called oils. If the double bond is *saturated* by hydrogenating the oil, a solidification ('hardening') occurs, an operation which is utilized in producing margarine by reducing vegetable oils with hydrogen. Conversely, a fat left exposed to the air undergoes oxidation along the fatty acid chain, resulting in double bonds and hence liquefaction, which is familiar as the first stage of the complex process whereby butter goes rancid. Butter is a mixture of animal fats, and the 'rancidification' is continued by a slight hydrolysis of some of the esters present, releasing the free acids, amongst them the foul-smelling butyric acid. Another familiar smell, that of burning fat, is provided by the heat-decomposition product of glycerol, *acrolein*.

The main function of the fats is, like that of glycogen and starch amongst the polysaccharides, to serve as a food store. Surplus food taken in by the animal is laid down as fat (adipose) tissue to be used again when harder times come. Such subcutaneous fat has other functions, too – it serves as an insulator against heat loss by the body, and also as a cushion for such delicate internal organs as the kidneys, which are generously embedded in fat. *Waxes*, which are lipid esters which contain rather longer-chain acids and alcohols than the fats, often have a role also as external protection – e.g. against water loss from the surfaces of insects, fruits, leaves, or petals.

Phospholipids

So far we have discussed neutral fats, but often a monoglyceride, instead of combining with other fatty acids, will react with the strongly electronegative phosphate group to give either monoacyl or diacyl phospholipids (depending on the number of fatty acids present). It is the reactivity of the phosphate group that enables phospholipid aggregates to form complex structures with other molecules, most notably with proteins in membranes. How the lipid aggregates form higher order structures depends on several variables: the degree of fatty acid substitution in a phospholipid, the presence or absence of strongly polar terminal groups, the degree of saturability in the fatty acids and the relative proportions of lipid and water in the mixture are all important considerations.

Diacyl phospholipid molecules are extremely well fitted for their role as membrane constituents mainly because they exhibit directionality in their relationship to H_2O – they tend to aggregate into bilayers, their hydrophobic tails, composed of the long fatty acid chains, are held together by weak bonds, whilst the strongly polar phosphate heads face outwards, enabling them to interact with the water that surrounds the membranes. We shall consider further aspects of membrane structure in the next chapter.

Neutral fats behave differently at similar concentrations mainly because they lack the strongly polar head so that their behaviour is determined by their non-polar residue and in solution they tend to form small round fat droplets – it is this particular feature that makes them so suited as a food store. Monoacyl phospholipids like lysolecithin tend to behave more like the neutral fats – they prefer to form micelles even at quite low concentration. This means that the insertion of just a few lysolecithin molecules into a lipid bilayer will greatly disrupt its organized structure.

Steroids

Also included in the lipid group by reason of their solubility in organic solvents is a group of substances which seems at first sight quite different from the fats and phospholipids. These are the *steroids*, whose many members are produced by a series of minor modifications on a basic pattern of seventeen carbon atoms linked together as four interlocking carbon rings:

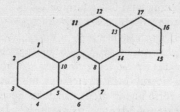

These rings, however, are simply the result of secondary folding of the long non-polar CH_2 chain, so the difference in structure is less great, though still important, than might at first appear.

A typical steroid is cholesterol, containing twenty-seven carbon atoms and an alcoholic group at carbon atom 3 of the steroid nucleus:

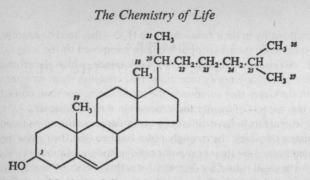

Other steroids derived from cholesterol include certain *vitamins* (e.g. Vitamin D), several *drugs* and *poisons*, the *bile acids* of digestion, and many *hormones*, in particular the sex hormones (see page 233). Derivatives of cholesterol itself circulate in the bloodstream and form part of the walls of arteries and veins, and disorders in the body's use of cholesterol are widely believed to be associated with arterial diseases, such as coronary thrombosis.

This completes our survey of the large molecules of which living organisms are made; it is time to turn from them to a discussion of the structures which, in life, they compose and with which they interact – the internal organization of the unit of life, the cell.

CHAPTER 4

THE ORGANIZATION OF THE CELL

So far, we have treated living tissue as if it were merely to be regarded as a convenient source from which to extract various classes of chemical, using greater or lesser degrees of violence, but in any event, without regard for the structures from which those chemicals have been derived. There are three ways of considering the cell; the first is the chemists'. For them, it is merely the logical extension of those chemicals we have so far discussed; a hierarchy runs from atom to molecule, from molecule to macromolecule with its primary, secondary, and tertiary structures. The cell is merely a higher order of these, its components may be regarded as quaternary structures, delicately related ensembles of proteins, lipids, polysaccharides, and nucleic acids, held in particular configurations by an appropriate ionic environment of water, sodium, potassium, and chloride ions. Implicit in this understanding (and we will return to this) is the chemist's belief that the higher order structures and their interactions are essentially predictable from the lower order ones; that primary structure determines tertiary which in its turn determines cell structure.

The second way of looking at the cell is the physiologists'; for them, like other 'organismic' biologists, the cell is the smallest unit of which tissues and organs are composed – the lowest level of a new hierarchy; it is the interaction of cells which is of interest rather than what goes on inside them.

The third viewpoint is that of the biochemist or cell biologist, concerned with both what the cell is made of and how it is maintained; where the chemist's quaternary structures and the physiologist's smallest living unit meet. This is the terrain of the present chapter.

In fact, in its earlier years, biochemistry got on by ignoring the cell as we have done in the previous three chapters. The reasons were largely methodological. Until the development of certain new tools of research in the 1940s – the ultra-centrifuge and electron microscope – biochemists had no means of relating observations about the behaviour of chemicals within the cell to those of biologists, who for many years had been using microscopes to map out the cell as a complex structure composed of many different subunits (or *organelles*, as they came to be generally called). But the time has now come when we can legitimately stop pretending that the cell is made of smooth paste like butter in a butter-dish, and can recognize that it is more like marmalade – full of small, solid lumps, often quite different in texture and composition from the surrounding jelly.

The sciences long concerned with the microscopic study of the cell are *histology* and *cytology*; their relation to biochemistry is confirmed by the two new subjects *histochemistry* and *cytochemistry*. The techniques of histology and cytology, developed over the last two hundred years, involve cutting very thin sections of biological material, mounting them on a microscope slide, staining them with a variety of dyes that colour different parts of the specimen more or less intensely, examining them at various magnifications, and finally attempting to relate back the two-dimensional picture seen down the microscope tube to the original three-dimensional living tissue. By such techniques the early microscopists were able to study plant and animal tissues and to show that they were divided into a set of box-like compartments separated one from the other by thin walls. The organ, or the body, they realized, was built of these boxes just as a wall is built of a set of individual bricks. They called the boxes, cells, and the thin walls, cell membranes (though plant cell membranes are themselves surrounded by a second line of fortification in the form of a more rigid *cell wall* of cellulose).

The cell was not merely the smallest unit of living organisms like plants and animals; for many forms of life it was found to be the *only* unit.

Moulds, such as yeast and *bacteria*, all manage to compress the whole range of their activities within the walls of one tiny cell – often considerably smaller, even, than the less-versatile plant or animal cells. Another familiar example of a one-celled animal is the little pond-water creature studied at one time or another by all biology students with access to a microscope – the *amoeba*. Early microscopists suspected, but could not prove, the existence of even smaller organisms, the viruses, the detailed study of which depended on the advent of the electron microscope in the mid-twentieth century. The virus completed the range of 'living units', and is something of a special case, as we will see later.

But average cells, from animals or plants, are all rather similar in both size and construction. They are likely to be about one-thousandth of a centimetre in diameter, and to weigh up to about one ten-millionth of a gram. In a person weighing about 80 kilos (11 stone), there might be as many as 100,000,000,000 (or 10^{11}) cells. Table 5 gives an idea of the range of sizes and weights with which one is dealing here. The cells from different species, say rats and humans, or algae and roses, differ from one another slightly in their properties, and so do the cells from the different organs of any one species – thus liver cells are slightly different from kidney cells and both from brain or blood cells.

But the differences, as we shall see, are not nearly as great as the similarities, and for much of this book we shall be describing the biochemistry of an 'ideal' or 'typical' animal cell, which might be that of most organs of most species. Only very much later will we come to consider in any detail how the biochemistry of the cells of different organs and species actually differ from one another.

According to the first theories, all cells were filled with a clear, jelly-like substance which was endowed with all the mystic properties of life, and was called *protoplasm*. By the 1830s, it had become clear that all cells contained at their centre a large, oval body, darker than the surrounding protoplasm and occupying as much as one-third of the total cell area. This body was called the *nucleus* (and at a later date, of course, gave its name to the nucleic acids – see page 67). The nucleus was separated from the rest of the cell by a thin skin of its own – the nuclear membrane. By the 1880s and 1890s, the microscopists had shown that the protoplasm surrounding the nucleus was itself far from being a clear jelly. Instead, it was lumpy and granular, filled with small specks of material which the highest-powered microscopes revealed as taking characteristic shapes as minute ovals, rods, and spheres. These specks were called *mitochondria*.

But even the highest powered light microscope can enlarge only up to about 3,000 times, and objects half-seen or only suspected

TABLE 5. *Dimensions of molecules and cells*

Species	Weight (gm.)	Atomic Weight	Length (longest axis) cm.
Hydrogen atom (H)	$1 \cdot 6 \times 10^{-24}$	1	1×10^{-8}
Glycine molecule (CH_2NH_2COOH)	$1 \cdot 2 \times 10^{-22}$	75	2×10^{-8}
Typical protein molecule	$3 \cdot 2 \times 10^{-20} - 1 \cdot 6 \times 10^{-18}$	$2 \times 10^4 - 1 \times 10^6$	$5 - 20 \times 10^{-7}$
Typical virus	$1 \times 10^{-16} - 2 \times 10^{-15}$	$4 \times 10^7 - 1 \times 10^9$	5×10^{-6}
Typical bacterial cell	8×10^{-7}		1×10^{-4}
Typical animal cell	$1 - 10 \times 10^{-5}$		$1 - 5 \times 10^{-3}$
Human	80,000		200

at this magnification could not be identified until, in the last years of the 1940s, the ordinary microscope could be supplemented by a new sort which used not light at all but a beam of electrons to examine specimens, and which could produce magnifications ranging from 10,000 up to one million or more. At once, a whole world of substructures unsuspected by the light microscopist became visible. The nuclei and mitochondria could be examined in fine detail to show a complex internal organization of their own, and the remaining 'cytoplasm' (the term which by now had replaced 'protoplasm' for whatever clear jelly was still considered to exist within the cell) was seen to be criss-crossed with a rich network of twisting strands, membranes, and small groups of tiny linked particles, the *ribosomes*, till it looked like a lace tablemat. Today, the typical animal cell can be pictured as drawn in Figure 6 or as photographed in Plate 1.

Table 6 shows the various particles that help compose the cell; the mitochondria are fifty to one hundred times smaller than the nuclei and the ribosomes up to sixty times smaller than the mitochondria. Very little remains of the pure protoplasmic jelly of 150 years ago in this modern picture.

An appropriate question for the biochemist to ask at this stage

FIGURE 6.

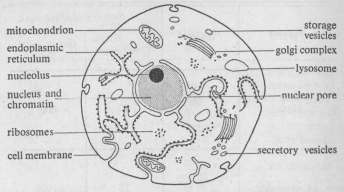

Schematic diagram of a 'typical' animal cell

is: what are the individual subcellular structures composed of and how do they maintain their structural individuality? Are there chemical differences and specializations which distinguish them?

It might be expected that certain activities are located in one organelle, others in another, and this localization is significant to an understanding of how the cell co-ordinates and regulates its activity. But before the biochemist can test these hypotheses, there must be available some method of separating the organelles from one another, and obtaining on a relatively large scale (which means, for the biochemist, quantities in the order of tens of milligrams of material) purified preparations of nuclei, mitochondria, ribosomes and so forth.

It is in the development of such methods that the techniques of *centrifugation* have proved invaluable. If the cell membrane is gently ruptured, all the internal organelles are released into suspension, like holding a cellophane bag full of sand and water under water and puncturing it. Now, if the resulting suspension of sand is shaken up and allowed to settle, the densest and heaviest grains fall to the bottom first, then the medium ones, whilst some very fine grains remain suspended in the water even after several hours. The rate at which the particles fall depends on several factors: their weight, their *density* compared to the surrounding

TABLE 6. *Sedimentation of subcellular particles in a centrifuge*

Particle	Shape	Sedimented after:	
		Time (mins.)	Gravitational g force
Nuclei		10	800
Mitochondria		15	12,000
Lysosomes		15	25,000
Ribosomes membranes		60	105,000

solution (a particle of cork, even a very large one, would not fall to the bottom however long one waited, as the density of cork is less than that of water, and it floats), and finally the gravitational force to which they are subjected. In our experiment with sand and water, the gravitational force is constant whilst the beaker holding the suspension is standing still, and the particles may take several hours to sediment. But ways exist of increasing the force towards the bottom of the beaker. Accelerating it upwards in a lift or rocket would be one way. Another would be to whirl it round very fast at the end of a long arm, when centrifugal force will tend to drive all the particles towards the part of the beaker furthest from the centre of the circle round which it is being rotated. Such a principle, called centrifugation, is used to simulate the increased force of gravity during rocket acceleration when training fledgling spacemen. It is also the principle of the spin driers attached to most modern washing machines.

In the hands of the biochemists, centrifugation is a technique enabling them to separate out particles from a suspension of biological material rapidly and efficiently. Modern laboratory centrifuges are capable of spinning tubes containing nearly a litre of suspension at speeds of 60 to 70,000 r.p.m., which is the equivalent of applying a force of 500,000 times gravity to the particles in the suspension. By employing a carefully selected range of speeds and times, it is possible to sediment and remove, first the cell nuclei, the heaviest particles, then the mitochondria, the next heaviest, and finally the ribosomes and membranes. The clear solution which remains after all the particles have been removed represents the *soluble* or *cytoplasmic* fraction of the cell. One such selection of speeds and times is shown in Table 6. A refinement of this method is density gradient centrifugation. Here the suspension is layered on top of a series of sucrose solutions of varying concentrations so that the density of the liquid increases from the top to the bottom of the centrifuge tube. During centrifugation the cell organelles move through the gradient until they reach the sucrose concentration that has a density equal to their own, where they will pile up (see Figure 7). So at the end of the run,

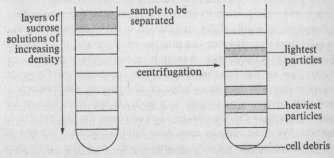

FIGURE 7. *Sucrose density of gradient centrifugation*

distinct bands can be seen at intervals down the tube, each band being a fairly homogeneous sample of one particular organelle. The accuracy and resolving power of this method is remarkable. Particles of very similar size can be easily separated and it can be done with just one spin instead of the many necessary with the differential method.

With this by way of background, we can now turn to a discussion of the individual subcellular components that can be seen under the electron microscope, isolated with the centrifuge, and studied chemically and biochemically.

MEMBRANES

Perhaps the best place to begin with our description of the cell is at its boundary, its point of contact with its environment, the cell membrane. The membrane has two main functions to perform. It has to maintain the structural integrity of the cell by its strength, insolubility and protective nature and it has to act as a selective barrier to substances in the external environment, allowing the passage of some but not of others. The membrane has the power to discriminate between harmful substances that would damage the cell if they gained entry, and therefore to prevent their entry, and substances that are essential nutritional requirements and must be allowed in to provide the cell with energy.

The membrane is also responsible for helping maintain the correct internal ion concentration so that such variables as pH, electric potential (the different concentrations and charges on ions inside and outside the cell) and osmotic pressure remain constant, often against wildly different external ionic concentrations that would tend to unbalance the internal environment if it were not for the membrane barrier. Research into membrane properties is today one of the most active areas in biochemistry. However, despite this activity, knowledge of the precise mechanisms of selective permeability is still very poor. It has been known for some time that the membrane is predominantly lipoprotein in nature. We have already seen how lipid molecules, because of their hydrophobic nature, spontaneously organize themselves into a double layer or sandwich when placed in a watery environment; their long non-polar tails line up together, forming the cheese inside the sandwich, whilst their polar heads face outwards, where they can readily bind to protein or glycoprotein components which represent the bread. The most widely accepted current theory of membrane structure is that of a core of lipid molecules arranged in a bilayer with its surface more or less covered with protein molecules, some of which penetrate right through the bilayer (see Figure 8). Because its secondary and tertiary structure is maintained by many weak bonds the lipid bilayer is very fluid and flexible; although again, because of the sheer number of internal hydrophobic bonds, it is at the same time strong and resistant to solvents.

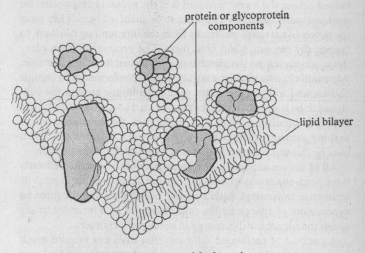

protein or glycoprotein components

lipid bilayer

FIGURE 8. *Fluid mosaic model of membrane structure*

There seem to be several ways for substances to enter the cell across the membrane. The simplest method, that works for the smaller molecules and ions, is one of passive diffusion, and occurs whenever their relative concentrations inside and outside the cell are different. Thus a chemical gradient in the case of non-electrolytes or an electrochemical gradient in the case of charged particles will exist across the membrane. Molecules will move across the membrane from the region of high concentration to the region of low concentration until an equilibrium is reached, when diffusion will cease. Diffusion is easier the more soluble the substance is in the lipid of the membrane, and may be further facilitated by the presence of holes or pores in the membrane, although this point is at present arousing a great deal of controversy in membrane research.

Active transport, on the other hand, is the process by which cells accumulate substances against a concentration gradient when the internal concentration of the molecules is already quite high and the external concentration is low. The molecule must then be helped across the membrane and it is the protein components on the membrane surface which seem to be involved here. They have the power to scavenge molecules from the surrounding medium, to specifically recognize and bind them. The protein will then carry the substance across the membrane and deposit it on the other side. Alternatively, the protein may undergo a conformational change and expand to span the width of the membrane so that the substance to be transported can be carried on a sort of protein bridge. The first binding protein may even pass the substance on to another, carrier protein, which is free within the membrane, and it may be this second protein that is responsible for translocation.

All of these mechanisms are made possible by the marvellously fluid lipid environment which does not restrict either lateral or transverse movement within it, but nonetheless they require an expenditure of energy by the cell. There will be a lot more to say about the source for this energy in subsequent chapters.

A method of facilitated diffusion that does not require much energy input makes use of the mechanism for transfer of Na^+ and K^+ ions that seems to exist within most membranes, the so-called ion pump. All living cells are polarized; the inside is electronegative with respect to the outside and this polarization is maintained by an unequal distribution of Na^+ and K^+ ions across the membrane. The importance of this polarization, too, will become clear in later chapters. The membrane pump is constantly pumping Na^+ ions out and K^+ ions into the cell so as to maintain a very pronounced ion gradient between the inside and outside of the cell. If a carrier protein in the membrane, together with its passenger molecule, complexes with a Na^+ ion, then the whole structure will move through the membrane by diffusion down the Na^+ gradient. Once inside the cell, the complex dissociates and Na^+ is expelled once more.

The membrane structure we have just described is not, however, confined to the cell boundary. In electron micrographs the whole of the cytoplasm seems to be interlaced with a series of double membranes forming a system of interconnecting channels (see Plate 1). This is called the endoplasmic reticulum and may be responsible for the transport of material from one part of the cell to another, although some researchers regard the structures as an artefact, a spurious product of the drastic methods of preparing tissue for the electron microscope. Their reasons for scepticism mainly derive from the fact that living cytoplasm is continuously moving and flowing within the cell and the existence of a complex membranous mesh would certainly impede this flow.

Also within the cytoplasm and visible in the electron microscope and separable by appropriate centrifugation procedures, is a system of disc-shaped, membrane-bound channels, three to twelve in number, stacked together to form the Golgi bodies, so called after an Italian microscopist of the late nineteenth century. These Golgi bodies are especially well developed in cells which are specialized for the production and secretion of substances like hormones, and they may be involved in packaging substances prior to extrusion from the cell.

NUCLEI

Of the organelles which can be separated by the fractionation techniques, the largest and heaviest are the nuclei. They are found in all animal and plant cells (with the exception of the red blood corpuscles of some mammals) and are intimately concerned with the reproductive mechanisms of the cell. They contain virtually all the cell DNA, a fair amount of RNA, and considerable amounts of basic proteins (histones, see page 68) that we have already referred to as being associated with DNA. The complex of DNA/RNA/protein is called chromatin and at certain stages in the growth cycle of the cell, becomes organized into well-defined *chromosomes*, whose role will become clear later.

Very often found within the nucleus is a small densely staining body called the nucleolus; the number of these vary in different cell types; they are full of a particular class of RNA. Some RNA is also present in the nucleus in the form of ribosomes, although these are mainly to be found in the cytoplasm (see below). The whole of the nucleus is constrained by a well-defined boundary, the nuclear membrane, which has the same type of lipoprotein bilayer structure as the cell membrane, except that it is perforated by pores whose role is obscure but which are believed to allow the passage of RNA into the cytoplasm (Plate 2). The significance of many of these structures and arrangements will become clearer when we discuss the mechanisms of protein synthesis in Chapter 10.

MITOCHONDRIA

The next in order of magnitude of the cell substructures are the mitochondria (singular – 'mitochondrion'). In many ways they are biochemically the most interesting, and have been intensively studied over several years. Each cell may contain up to a hundred or so mitochondria, each egg-shaped and between 1 and 4μm in length (a micrometre is 10^{-6} metres, or a thousandth of a milli-metre). Each mitochondrion is surrounded by a double mem-brane, an inner and an outer one, each in turn being composed of the by now familiar lipid bilayer. The outer membrane is smooth, but the inner is pulled into long folds that run across the entire width of the mitochondrion, so that in section the mitochondrion appears in the electron microscope like a pile of custard-cream biscuits. Each of the biscuits represents a fold of the inner mem-brane running across the mitochondrion, whilst between each fold is the custard cream, the concentrated solution with which the mitochondrion is filled. The folds of the inner membrane are known as *cristae* (Figure 9, Plate 3).

In part, this membrane structure serves the same purpose for the mitochondrion as does the cell membrane for the cytoplasm, that is, of regulating the traffic of chemicals between inside and outside, but there is more to it than that. The mitochondrion, as we shall see in succeeding chapters, is the powerhouse of the cell, involved in the mechanism of the production of utilizable energy by the breakdown of substances derived from food. Each of the different regions of the mitochondria, inner and outer membranes and the liquid intercristal space, has its part to play in this integrated process, and the structure of the mitochondrion is a precise expression of its function.

FIGURE 9. *A mitochondrion*

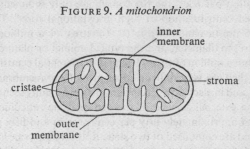

LYSOSOMES

Next in size to the mitochondria comes a group of particles which, discovered in liver by Christian de Duve, of Brussels, in 1953, provided a happy solution to an interesting 'teaser' which had for long intrigued biochemists. The cell, when broken up by homogenization, contains a variety of enzymes (proteins, whose function in assisting chemical reactions to occur we will discuss in the next chapter) which, between them, are capable of destroying nearly all the substances which go to make up the cell. How are they kept in check by the living cell and prevented from running amok? De Duve demonstrated that they were all contained within a class of particles he called lysosomes (Plate 4).

Under normal circumstances, the lysosome is surrounded by a membrane which cages the enzymes in, keeping them from attacking other cell components. However, if the cell is damaged or dies, the lysosome membrane bursts, releasing the enzymes into solution and hence bringing about the rapid total dissolution of the cell.

The lysosomes thus function as a party of cellular hyenas, harmless to the healthy, but quickly destroying sick or injured cells. Since de Duve found lysosomes in liver, they have been demonstrated in many, but not all, other animal organs, and it is sure that they have a general scavenger function.

RIBOSOMES

The smallest particles, the ribosomes, are only sedimented after prolonged centrifugation at high gravitational force. They are about 15 nm (a nanometre is 10^{-9} metres, or a millionth of a millimetre) in diameter, and the typical animal or plant cell contains about a million of them. Using the differential centrifugation method they are very difficult to separate from membrane components, and in electron micrographs they are mainly seen studded along the external surface of the endoplasmic reticulum. They consist of approximately fifty per cent protein and fifty per cent RNA, and are composed of two structural subunits, one large and one small (see Figure 24, page 204). The main function of the ribosome is in the manufacture of protein molecules from their amino acid building blocks and each of the subunits plays a different part in this process. We will discuss this in detail later. They are often found in multiples called polysomes, especially in cells that are most active in protein synthesis.

In many ways the ribosome can be regarded as a prime example of a quaternary, higher order molecular structure, for just as lipids and proteins in appropriate conditions will orient themselves so as to form a bilayered membrane, so too will the component RNA and protein of the ribosome. If each ribosomal subparticle is carefully fractionated it can be separated into thirty or forty separate protein and RNA fractions. If these fractions are mixed together in approximately the right proportions, the conformational relationships between the individual RNA and protein molecules are such that they bind together and can be sedimented as reformed ribosomes once more, capable of protein synthesis. This self-assembly must be a fundamental process governing the organization and structure of cells and their organelles, but it is only now beginning to be studied.

The Organization of the Cell

CYTOPLASM

When all the particles have been sedimented out of suspension, there is left a clear solution which contains all those substances which, in the intact cell, are not allotted to any of the membranes or particles but are presumed to remain free in solution in those spaces still left between the various particles and the reticulum. The solution contains many of the simpler low-molecular-weight constituents of the cell, such as sodium, potassium, and phosphate ions, and also a number of enzyme systems with which much of the rest of this book will be concerned.

Even from this brief account, and before we really begin to come to grips with the nature of the various biochemical functions for which they are responsible, we can see that it is possible to regard the cell as a series of more or less self-contained compartments, each containing particular small and large molecular weight components and involved in interlocking biochemical reactions. As we now come to consider these reactions in more detail, moving from the rather static picture of chemicals and cell structures – biochemistry as analysis – to the dynamics of biochemistry in action, it will become clear that the high level of structural organization has evolved as a most efficient way of ensuring that the cell can regulate its own internal biochemistry in the interest of smooth and efficient running.

CHAPTER 5

WORK AND ENZYMES

The biochemical world we have until now been describing is a static one; it is a world of things rather than events, of photographs rather than cinefilms. In order to obtain this picture, we have had deliberately to destroy the living tissue at some arbitrary time, and systematically to fragment it into its component chemicals. Yet by doing so we have ignored just those characteristics which make life different from death. The proteins, nucleic acids, lipids, carbohydrates, of a living animal are to all intents and purposes identical with those of one just dead. But no one would mistake the one for the other. In this chemical world we have entered, in which our bodies have been reduced to chemical symbols, where has the life gone? Can we describe the difference that we know exists between life and non-life in chemical terms?

Of course, we can. But to do so we must shift the perspective from which we have been studying the chemistry of the cell. We must no longer ask what they are made of but, rather, how are they kept in existence? For the highly organized macromolecules we have been describing all have one most important characteristic in common. By comparison with the simpler molecules they are all highly unstable. They readily begin to break down as soon as their environment alters beyond certain well-defined limits of pH and temperature. Thus, in the living system, any change that tends to alter these variables beyond their permissible limits must be opposed. We observed before how solutions of amino acids tended to resist attempts to change their pH by the addition of acid or alkali. The amino acid solution acted as a *buffer* against pH change. Similarly, the living organism is a mechanism which operates as a *buffer* against changes in its environment (we have already remarked on one of the buffer mechanisms, the selectivity of the cell membrane, and we will come across many more as we continue). When simple chemical systems operate to oppose change, they do so according to the principle of a French chemist, Le Chatelier; it was the French physiologist, Claude Bernard, who recognized that the same process (now known as homeostasis) also occurred in living organisms.

The second characteristic implicit in the macromolecules is that they are all extremely *unlikely* substances. Those materials which life produces in such abundance still defeat the synthetic techniques of the chemist. In the living cell, such molecules cannot arise purely by random chemical reactions; they must be synthesized according to precisely planned pathways which can achieve a specificity far beyond that of the chemist. There must be mechanisms within the cell which can distinguish between even such close relatives as the (+) and (−) isomers of amino acids, or between sugars such as glucose and galactose.

This second problem would not perhaps be so overwhelming if the synthesis were, so to speak, a once-for-all job, if the lipid, protein, and carbohydrate molecules only had to be made in the desired quantity, laid down in the appropriate structure, and then were able to go on functioning indefinitely, until 'fair-wear-and-tear' demanded their replacement. But such is not the case. The now classical application of radio-isotopes to biochemistry, pioneered in the 1920s by Hevesy in Copenhagen and continued with great distinction by Schoenheimer in New York, revealed that all body-components were in a constant state of flux; that protein, lipid, and nucleic acid molecules were constantly being renewed, old molecules being broken down and new ones synthesized to take their place. Even the molecules of such stable and unreactive tissue as bone and cartilage, which used to be regarded as quite inert, were found to have quite a short life-expectancy. Throughout the body, the molecule which survived for more than a few days without undergoing change was found to be the exception rather than the rule. The discovery of the prodigality of this constant flux of molecules, which leaves even the built-in obsolescence of such mass-produced items as cars and television sets standing, revolutionized biochemical thinking. It became clear that one, perhaps *the*, major function of the living cell was the constant re-creation of itself from within.

So before it can even begin to act on its external environment, the cell, or the living animal, has to provide the mechanisms whereby, first, it can protect itself against dissolution and destruction by the outside world, and, second, it can continuously resynthesize its more complex parts from much simpler molecules. Two problems between them embrace that part of biochemistry concerned with kinetics:

(1) How does the cell set about achieving the large number of highly complex chemical reactions necessary for the synthesis of macromolecules? And

(2) How does it obtain the energy required to accomplish its varied tasks? For it is clear that the activities we have been describing represent *work*. To collect together several hundred amino acids at a precise point, to join them by means of peptide bonds, to coil and fold the resultant molecule, and to dispatch it at last to its appropriate site within the cell means that work has to be done upon the molecule. Similarly, to remove or render harmless poisonous or dangerous substances, and to keep the cellular pH within its desirable limits, requires work. And, of course, the moment the cell begins acting on its environment – the muscle cell by contracting, the nerve by transmitting impulses, the hormone-making cells by secreting their hormones – more work is done.

I. WORK AND ENERGY

The terms 'work' and 'energy' used in this sense may perhaps be unfamiliar ones. The concept of physical work is straightforward; we do work when we lift a weight from the ground, and the measure of the work done is represented by the mass we have lifted times the vertical distance through which it has travelled. Equally, the weight, once lifted, possesses potential energy;* when released, it will fall to the ground, and, in falling, can perform work in its turn – it could be coupled to another weight over a pulley, for example, or used to drive a grandfather clock. Also, in physical terms, we are well aware of the law of the conservation of energy – we can get no more work out of a system than we have initially put into it. The grandfather clock will go on ticking

* Strictly, it is not the weight that possesses the energy, but the gravitational field acting upon it. When the weight is lifted, this energy is 'stored' in the gravitational field, which 'gives it back' when the weight falls.

until the weight has fallen again and all its potential energy has thus been used up. Then the clock will stop, and will only start again when the weight is lifted once more and hence more potential energy is pumped into the system.

This law of the conservation of energy makes it easy to understand the concept of work in a physical system, but it is not immediately obvious that work and energy are factors in chemical reactions too. Consider, however, the combination of carbon and oxygen to give carbon dioxide. We write the equation

$$C + O_2 \rightarrow CO_2$$

but, in doing so, have omitted one important fact about it – that, during its course, about 420,000 joules* are released for every twelve grams† of carbon burned. In fact, the equation ought more accurately to take the form

$$C + O_2 \rightarrow CO_2 + 420 \text{ kJ}$$

We can use these joules as a source of warmth, in a coal fire, or can convert them into other forms of energy, as when the coal is burned in a steam engine or an electricity generator. Thus the energy released in this chemical system is directly convertible to the energy we earlier described in a physical system – we could replace the falling weight of the grandfather clock, for instance, by driving the clock with electricity generated by burning coal. Equally, the law of conservation of energy holds in the chemical as well as in the physical system. When carbon reacts with oxygen the system releases energy, and in doing so itself loses energy. In order to convert the CO_2 back into $C + O_2$ again it would be necessary to supply exactly the same number of joules as those lost when CO_2 is formed.

The reaction $C + O_2 \rightarrow CO_2$ is thus described as *exergonic*, or energy releasing, whilst the reverse reaction, $CO_2 \rightarrow C + O_2$ is

---

* A joule is a unit of energy used when dealing with heat; it is approximately the amount of heat required to raise the temperature of 1g of water through 0·24 °C; there are 4·18 joules in the – perhaps more familiar – calorie. The term kJ (kilojoule) is used for a unit equivalent to 1,000 joules.

† This is the 'gram-atomic weight' or the atomic weight of carbon (which is twelve) expressed in grams. Similarly, the gram-atomic weight of oxygen is sixteen grams and the gram-molecular weight of water is eighteen grams.

The Chemistry of Life

endergonic or energy-requiring. The energy change that occurs during such reactions, considered as occurring under standard conditions, is referred to by the measure ΔG^0, the standard free energy. In any reaction, if ΔG^0 is negative, the reaction is exergonic – that is, it can do work. If ΔG^0 is positive, the reaction requires energy input in order to take place.

Most of the reactions performed by the cell are endergonic; they require a source of energy. What is this source? Plants, as is well known, have a mechanism whereby they can tap the torrents of energy released every day by the complex atomic fusions which occur within the sun and pour down upon the earth as heat and light. The plants use the light energy for synthesizing sugars, such as glucose, and polysaccharides, such as starch, by the 'fixation' of CO_2. They can then break down the glucose and polysaccharides to release the energy once more when they require it. Animals have no primary energy source of this sort, and must rely on being able to find a ready-made supply of potted energy in the form of sugars. They achieve this by devouring the plants and using the materials which they have made. More sophisticated animals yet, of course, devour the animals which devour the plants, and the cycle of mutual interdependence is completed when the plants themselves utilize such waste-products as carbon dioxide which the animals release.

But the important thing about the whole cycle is that throughout it the law of conservation of energy holds. Many experiments have been done in which animals, or human volunteers, are kept for several days in closed boxes, whilst the energy provided for them in the form of food, and that utilized – as heat output, carbon dioxide production, nitrogen excretion, and work done (for example, the riding of stationary bicycles) – are measured. Within the closest obtainable experimental limits, both humans and animals obey the law of conservation of energy, input and output being exactly balanced.

As with the body, so with the cell; and, in describing the many chemical syntheses and activities of the cell, we must at the same time ask also from where the energy for the performance of this work has come.

The principal source of energy for the cell is, of course, glucose.

Work and Enzymes

The complete burning of glucose goes according to the equation

$$C_6H_{12}O_6 + 6O_2 \rightarrow 6CO_2 + 6H_2O + 2{,}820\,kJ$$

The cell performs this reaction, but it does so in a rather round-about manner. It breaks the glucose down by a series of reactions, each time releasing a small amount of energy only. The release of such a large amount of energy in one lump, so to speak, would be too much for the cell to cope with; much of the energy would be dissipated as heat, quite likely destroying the cell in the process. What the cell has been able to achieve for glucose is its controlled stepwise breakdown to provide a steady source of energy – just as an atomic power station is able to harness, control, and make useful the vast quantities of energy released as heat in an atomic explosion.

Glucose is converted to CO_2 by a process involving nearly thirty different steps in each one of which a small amount of energy is released. Thus the cell obtains its energy in the form of small 'packets' which can be carefully conserved and used systematically. But it is important to note that in the final analysis the total amount of energy released during the glucose breakdown is exactly the same whether it is burned directly to carbon dioxide or passed through any number of different intermediate processes on the way. By way of illustration, consider the reaction

$$V \rightarrow Z + 100{,}000\,j$$

Now suppose that the reaction instead occurs by way of a number of intermediates, W, X, Y. We can write the equations

$$
\begin{array}{llll}
V & \rightarrow & W + & 25{,}000\,j \\
W & \rightarrow & X + & 25{,}000\,j \\
X & \rightarrow & Y + & 25{,}000\,j \\
Y & \rightarrow & Z + & 25{,}000\,j \\
\hline
\end{array}
$$

Overall reaction $\quad V \rightarrow Z + 100{,}000\,j$

By whichever route V is converted to Z, the *total* energy change during the reactions is always the same. This is another consequence of the law of conservation of energy, and it means that the cell has nothing to lose, but a lot to gain, in its roundabout approach to free energy release.

We began by framing two questions with which to approach the world of biochemistry-as-kinetics: how does the cell achieve the chemical transformations that it undertakes, and where does it get the energy for so doing? We have now provided a general answer to the second question. But before we can proceed to answer it more fully, it is necessary to think again about the first.

II. ENZYMES

Few chemical reactions proceed completely spontaneously. Most need to be triggered off. Even thermodynamically favourable reactions, like the burning of carbon in oxygen that we considered in the last section, do not begin entirely by themselves. Carbon, as anyone knows who has tried to start a coal fire, is perfectly stable under normal conditions, and refuses to react with oxygen until its temperature is raised above a certain point. Once that point has been reached, the reaction goes ahead merrily until lack of carbon, or lack of oxygen, stops it. Until it has been reached, the carbon will sit indefinitely in the grate of the fire, refusing to burn. The amount of heat that has to be applied to the coal before it will catch fire represents the *activation energy* of the reaction. Once it is alight, the coal releases far more energy than had to be put into it to start it burning, but despite this, it would not begin without the added push. In this, the lump of coal waiting to catch light may be likened to a ball in a small hollow at the top of a long hill (see Figure 10).

In order to release the potential energy present in the ball by virtue of its position at the top of the hill, a small amount of activation energy, in the form of a push, must be applied to it so as to roll it out of the hollow and over the top of the hill. Chemists can supply activation energy to their reactants without difficulty, by warming them in a test-tube, increasing the pressure on them if they are gases, altering their pH if they are liquids. The cell can do none of these things. The chemicals of which it is made begin to disintegrate if their temperature is raised much above thirty-seven degrees Centigrade or if their pH is shifted much from 7.0. How is it to provide the activation energy necessary for reactions to begin?

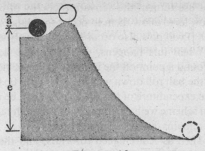

A small amount of activation energy 'a' is needed to get the ball out of the hollow. Once out, though, it rolls downhill releasing potential energy 'e'.

If all else failed, it could, of course, supply energy from its intro-cellular stores, but this would be a rather extravagant process. In some cases, as we shall find, that has to be done, for nothing else will suffice. But in most cases, a way round the difficulty has been found. What the cell does is to resort to low cunning rather than brute force. Another way of lifting the ball over the ridge of Figure 10 would be to go on rolling it along *behind* the ridge until a point was found where the ridge became sufficiently low for only the tiniest push to be needed to set the ball rolling down the hill. Chemically, this means that if we are trying to perform the reaction

$$A \rightarrow B \qquad (1)$$

which, although thermodynamically favourable, still requires a fairly high activation energy, we may instead look for another reaction, involving an intermediate X, which can react according to the equations

$$A + X \rightarrow AX \qquad (2)$$
$$AX \rightarrow B + X \qquad (3)$$
$$\overline{\qquad\qquad\qquad\qquad}$$

SUM $\qquad A \rightarrow B \qquad (4)$

of which the sum, reaction (4), is identical with the reaction (1) in which we were originally interested.

X is only of use to us if the activation energy of reactions (2) and (3) is so low as to allow them to occur spontaneously. In that case, reaction (4) will now also occur spontaneously as a result of (2) and (3). When this happens, we shall have achieved our purpose and found a point on the hilltop where the ridge was low enough to let the ball roll down without having to be heaved over the edge by the expenditure of much energy. It will be noticed that in the reaction scheme we have drawn, X, although taking part in the reactions, remains, at the end, unaltered.

Substances such as X are called *catalysts*, and the process of lowering the activation energy of a reaction is *catalysis*. Catalysis is a pretty common phenomenon in chemistry and finds much application, both in the laboratory and in commerce. 'Catalytic cracking' of oil to give petrol and other derivatives, or the use of 'Raney nickel' as a catalyst in the reduction of oils, saturating their double bonds and turning them into fats such as margarine, are well-known processes.

Enzymes

Catalysis is practically universal in biochemistry, and the class of biochemical catalysts are collectively known as *enzymes*. We saw in the Introduction how they got their name when the brothers Buchner first extracted a preparation of yeast by grinding it with sand in a mortar and preparing a juice which could ferment sugar catalytically to give alcohol. A number of general rules govern the behaviour of enzymes, as of all catalysts. The sum of the rules is that enzymes can do only certain well-defined jobs, and can never violate the conventional laws of chemistry and physics, even though at first sight their incredible versatility might seem almost inexplicable.

(1) A catalyst can never catalyse a thermodynamically un-favourable reaction. It can never change an impossible into a possible reaction, but only make a possible but difficult one rather easier. It can never roll the ball uphill again once it has rolled to the bottom of the hill.

(2) A catalyst can never change the course of a reaction. If the reaction in the absence of the catalyst is A → B, all that the cata-

lyst will do is make A → B easier. It won't make A → C happen instead.

(3) Nor can a catalyst change the *equilibrium* of a reaction. In a reversible reaction

$$A + B \rightleftharpoons C + D$$

if the reactants are mixed together and left for long enough, ultimately a characteristic and quite precise ratio of (A + B) to (C + D) will be arrived at, whether one started one's reaction mixture with A and B only, with C and D only, or with a mixture of all four. This ratio will be the same whether the reaction is catalysed by an enzyme or left to reach equilibrium by itself.

(4) In certain circumstances, though, a catalyst might exert a *directing* influence over a reaction. Thus if the reactions

$$A \rightarrow B$$
and
$$A \rightarrow C$$

are *both* thermodynamically possible, a catalyst might help decide between them by lowering the activation energy required for one more than it lowers it for the other. This directive role can be very important amongst enzymes strategically placed at certain points along the pathways of metabolism followed within the cell.

(5) Finally, the general rule of catalysis, that although the catalyst may participate in the reaction it catalyses, in the end it is recovered unchanged from the reaction mixture. This is implicit in the equations (2) and (3) of page 101. It means that a very tiny amount of the catalyst can be used over and over again by a very large number of molecules of the reactant. Some enzymes, for example, are used and released again so rapidly that one molecule of an enzyme can catalyse the transformation of half a million molecules of the reactant per minute. Thus enzymes exert an influence on events within the cell quite out of proportion to the actual amount of enzyme present.

Enzymes, then, are the tools the cell uses to manipulate, cut up and stick together the molecular raw material with which it is presented. One set of enzymes takes glucose down its long pathway to carbon dioxide, another set synthesizes proteins from amino acids or fats from fatty acids. Enzymes have, however, a special characteristic that distinguishes them from other, non-biological catalysts. Whilst such catalysts as Raney nickel will readily catalyse a whole class of reactions of the same general type – that of reduction by hydrogen – enzymes are very highly selective about the reactions they choose to speed up and those they will not catalyse. They are not general-purpose but custom-built tools, and the cell requires a different enzyme for practically every reaction it carries out. This specificity is not entirely absolute; there are enzymes, such as esterases, which will catalyse the hydrolysis of an ester

$$R.COOR' + H_2O \rightleftharpoons R.COOH + R'OH$$

without much regard for the nature of the groups R, R', always provided there is an ester linkage between them.

Other enzymes are slightly more choosey, and demand that at least one half of the molecule they act on be determined; acid phosphatase, for example, catalyses the hydrolysis

$$R.O.PO_3H_2 + H_2O \rightleftharpoons R.OH + H_3PO_4$$

Again, R can be any one of many possible groups, but, if the phosphate half of the molecule is replaced by anything else, the enzyme at once refuses to work.

For the majority of enzymes, however, absolute specificity is the rule. For example, lactic dehydrogenase catalyses the oxidation of lactic to pyruvic acid

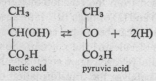

but does not, for example, act on lactic acid's near neighbour, hydroxybutyric acid

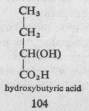

What is more, not only is lactic dehydrogenase specific for lactic and no other acid, but it also discriminates between the two optical isomers of the acid, only accepting one form. Most enzymes are tailor-made like this to catalyse one and only one reaction.

As the cell performs many different reactions, this means that there must be a great number of different enzymes in any one cell. And indeed this is true. Getting on for a thousand different enzymes have now been recognized, and more are being discovered all the time. A cynical enzymologist once remarked that given *any* reaction an organic chemist could suggest, it would be possible to find, somewhere in the world of living things, an enzyme to catalyse it. No one has yet proved this wrong, and, indeed, enzymes are beginning now to take their place amongst the tools of industrial synthetic chemistry. Many new substances, notably drugs and antibiotics, are produced with the help of enzymes. For, like any other catalyst, give an enzyme satisfactory working conditions and enough starting material to work with (the substance with which an enzyme reacts is called its *substrate*) and it will go on breaking down substrate and churning out reaction products practically indefinitely.

Not every cell in every species, of course, has all the enzymes that have been discovered. Many enzymes are quite specialized and are to be found only in particular species of bacteria or fungi. But most cells contain several hundred different enzymes, and these same enzymes will be found over and over again in mice and humans, streptococci and carnations, mushrooms and sharks, so universal are the reactions they are called upon to perform.

The first enzyme to be obtained purified and in crystalline form was urease (in 1926, by Sumner in America). Since then many hundreds have been extensively purified and isolated as crystals. All have turned out to be proteins, and it may be taken as a general rule that all enzymes are in fact largely protein in composition, although many also have non-protein material such as metal ions or nucleotides bound to them.

Enzyme kinetics

The kinetics of enzyme-catalysed reactions have been very intensively studied, not least because this is one of the few parts of biochemistry that can be brought under rigid mathematical scrutiny and analysis. Many of the characteristics of enzyme action are restatements of the general laws of catalytic behaviour that we have already summarized; most of the others can be deduced from the knowledge that enzymes are proteins.

Some of the key features of the behaviour of enzymes can be demonstrated simply by incubating an enzyme with its substrate and measuring the amount of product formed in a given time. From such a measurement, one can calculate the *rate* or velocity of the reaction, V (amount of product formed/time). If one performs this experiment several times with varying amounts of substrate, S, but the same amount of enzyme present, one can plot a graph of velocity versus substrate concentration (see Figure 11). Such a

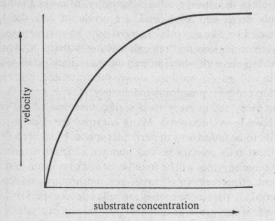

FIGURE 11. *The effect of substrate concentration on rate of reaction*

graph shows that as the substrate concentration increases, so does the velocity, at first, but soon levels off to a point at which it cannot be increased by any further addition of substrate – the maximum velocity, Vm, which occurs when there is enough substrate present to complex with all of the enzyme. An estimate of Vm is therefore also a rough indication of the amount of enzyme present in a particular preparation. However, termination of the reaction may also occur because the equilibrium-point of a reversible reaction has been reached and substrate and products are now present in balanced concentration, or because during the reaction there has been a change in pH, or the enzyme has been inactivated in some way.

For instance, the velocity of enzyme reactions is very much dependent on temperature. If the temperature is raised, the initial reaction velocity increases (as is the case for all chemical reactions, which go faster at higher temperatures), but also the enzyme is less stable and sooner inactivated. Most mammalian enzymes work best at thirty-five to forty degrees Centigrade, the temperature at which the body is normally maintained, whilst fish and other cold-blooded animals often have enzymes which function well at lower temperatures.

Their lability to heat is one indication of the protein nature of enzymes. Another is their sensitivity to pH. All enzymes have one pH at which they work fastest, and slight shifts in pH away from this point bring sharp falls in catalytic activity. A typical pH curve for an enzyme is that of Figure 12, and it is plain that this effect is

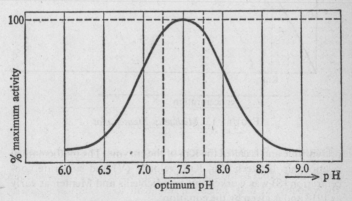

FIGURE 12. *Enzyme activity and pH*

related to the changes in protein structure that occur when the pH alters. For most enzymes, the optimum pH is not far from neutrality, pH 7.0, although some, like the amylase of saliva (which is alkaline), work best at somewhat more alkaline pHs.

Some enzymes have optima as high as 9 or 10, whilst others, such as the pepsin of stomach digestive juice, function at the high acidity of pH 2–3.

Hence in studying enzyme catalysed reactions, one rarely looks at maximum velocity, but tries instead to measure the initial velocity of the reaction, i.e. the reaction rate over the first part of the graph of Figure 12. If one looks at a substrate concentration that gives half-maximal velocity – $\frac{1}{2}$Vm – (shown in Figure 13), this concentration turns out to be a constant for any

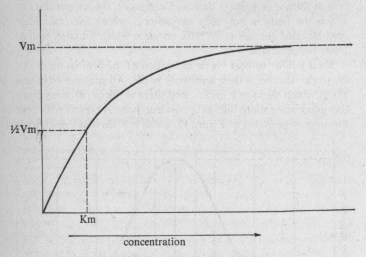

FIGURE 13. *Michaelis–Menten plot*

one enzyme and is called the Km of the enzyme. The mathematical relationship between Vm, Km, velocity (V) and substrate concentration (S) was worked out by Michaelis and Menten as early as 1913 and is given by the equation

$$V = \frac{Vm\,S}{Km+S}$$

Km is called the Michaelis constant and is a measure of the *affinity* of the enzyme for its substrate. The lower the Km, the higher the affinity. The important thing is that this equation holds true for most enzymes and the curve of the equation is that drawn in Figure 13. So any deviations from this shape when V/S is plotted will tend to indicate, for instance, that the enzyme is not pure or that it does not have a high specificity for its substrate. Although kinetic studies of enzymes and substrates studied in isolation from other cell constituents can only be an approximation of what is actually happening within the cell, the Michaelis-Menten equation, or derivations of it, are still extensively used (if only because the curves please the enzymologists!).

Mechanism of enzyme action

How do enzymes work? Like all catalysts, they are assumed to form a complex with their substrates which is then broken down to release the enzyme and the reaction products. Detailed study of the mechanisms of enzyme reactions, and, in particular, measurement of the velocity with which the enzyme converts substrate into products under changing conditions, and with varying amounts of reactants present, confirmed this prediction theoretically long before it could be tested experimentally. But, more recently, it has been possible to provide direct experimental proof of this complex formation. In some cases, for instance, where the substrate has characteristic wavelengths at which it absorbs or emits light (as do many organic substances), the formation and breakdown of the enzyme-substrate complex has been followed visually by watching the reaction and the changes in light absorption of the reactants in a spectrophotometer. Under certain circumstances, for a few enzymes, it is even possible to isolate the enzyme-substrate complexes and study them directly.

Knowing that a complex is formed between enzyme and substrate is one thing; discussing the precise details of the reaction of any one enzyme is another, especially as the enzymes themselves are proteins and their actual chemical structures largely unknown. But despite the huge size of the enzyme molecule, the reaction that takes place involves only a small part of the amino acid chain, perhaps only half a dozen or a dozen amino acids in length, although in some enzymes the rest of the chain may play a part in controlling the reaction. This critical region is called the active centre of the enzyme. We have seen how the strong sulphide bonds that form between cysteine residues as the amino acid chain twists into its secondary structure can hold parts of the chain in a definite and characteristic configuration. Thus pockets or clefts of a particular shape may be buried deep within the protein molecule (as in Figure 14, page 111), protected from the environment of the rest of the cell and therefore able to control their own pH conditions, in turn affecting the electrical charges carried by the amino acids in the region. The strength of union between enzyme and substrate is due to the formation of a large number of weak bonds, made possible by differences in electric charge distribution. Before these opposing charges can attract each other they have to be extremely close together. Thus the shape of the molecules at the point of attachment has to be exactly complementary to enable sufficient bonds to form. Any small mismatches will diminish the number of bonds that can be formed and the molecules will fail to recognize each other.

Thus the substrate fits into the enzyme like a key into a lock in such a way that the enzyme-substrate complex is much more reactive than the substrate alone. The enzyme may exert this effect in several ways. The substrate molecule may be distorted or strained as it becomes bound to the enzyme, so making it easier to split; or, because of the precision of interaction, the enzyme may hold the substrate in exactly the right orientation for it to react with a second substrate molecule. The enzyme may also be responsible for concentrating substrate molecules in a localized area, therefore making it much more likely that they will meet and react with each other. This is especially true of the many enzymes that are firmly attached to membranes rather than freely wandering within the cytoplasm. As soon as the substrate has been transformed into the reaction products, they no longer fit into the space on the enzyme surface and are released into the surrounding solution, freeing the active centre for another substrate molecule.

In tailoring a shape to fit the substrate, the protein chain is frequently not adaptable enough, and the enzyme makes use of other substances to complete the pattern. Metal ions are amongst the most frequently used, particularly the ions of magnesium and manganese, both of which contain two positive charges and can thus provide a useful link between protein chain and

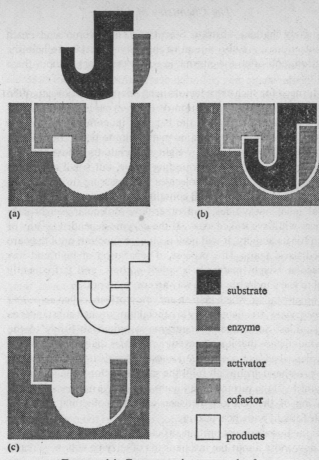

substoate

substrate

enzyme

activator

cofactor

products

(c)

FIGURE 14. *Enzymes, substrates, and cofactors*

(a) *Before reaction*
Substrate and enzyme plus cofactors come together.
(b) *During reaction*
All components locked together, thus activating substrate molecule.
(c) *After reaction*
Broken substrate molecule is released and enzyme is ready for further reaction.

negatively charged substrate. Sometimes entire compounds, such as nucleotides, are also bound to the enzyme surface to help the reaction, and often the enzyme is powerless to act without these additions.

The need for such extra factors in an enzyme reaction can often be shown by putting a solution of the enzyme in a sealed cellophane bag and suspending the bag and its contents in distilled water. The cellophane contains many minute pores, big enough to allow all low molecular weight materials to diffuse through them and out into the surrounding water, but small enough to prevent the large protein molecules from passing through. Thus after a few hours the bag will contain only the protein, and all the metal ions, nucleotides, and other low molecular weight substances will have leaked away. If the enzyme depended on any of them for its activity, it will now no longer function until they are added back again. This process of separation of high and low molecular weight materials is called *dialysis*, and is frequently used to learn more about the way an enzyme works.

The substances which help an enzyme work are called *activators* or *coenzymes*; frequently they are metal ions or such substances as nucleotides. Sometimes a coenzyme is bound so firmly to the enzyme surface that it is not even removed by dialysis; in this case it is described as a *prosthetic group*. But activators, coenzymes, and prosthetic groups all fulfil the same function of tailoring the enzyme to fit the substrate. In Figure 14 we have made a schematic drawing of the relations between enzyme, cofactors, and substrate for a hypothetical enzyme.

As the knowledge of protein structure increases, we may expect to know more about the mechanisms of enzyme action. Already information on the amino acid sequences around the active centres of quite a few enzymes has been obtained, and it has been observed that there are striking similarities amongst the active centres of various different enzymes within the same general class, such as phosphatases and esterases. Most notably, nearly all of these enzymes seem to possess at their active centres one or more serines whilst other classes of enzyme make use of the sulphur-containing amino acid cysteine, suggesting that —SH groups play a role in complex formation in these cases. It may well be that within the next few years certain general rules about enzyme mechanisms will become clear.

Enzyme inhibition

The extreme precision of the chemical tailoring at the active centre of the enzyme protein means that anything which alters the shape of the molecule at this point will interfere with the formation of the substrate complex. Enzymes which depend on a particular distribution of electric charge at the active centre will rapidly be inactivated by changes in pH which alter this distribution. Enzymes which depend on sulphydryl groups will be inactivated by any substance which reacts with SH groups (as do salts of lead or mercury, for instance). Enzymes that depend on metal ion activation will cease to work when their activating ions are replaced by other metals, such as copper or iron. Substances which inactivate enzymes in this way are called *inhibitors*, and enzyme inhibition may or may not be reversible depending on the particular circumstances. If the process is irreversible, the enzyme is said to be poisoned.

Another form of inhibition is indicated when we remember that not only has the enzyme to be tailor-made to fit the substrate, but that the substrate itself must also precisely fit into the space available for it on the enzyme surface. If a large alteration in the substrate molecule occurs, it will no longer fit the space left for it and no reaction will occur. But if it is only slightly changed, it may still be sufficiently like the original molecule to slip into the active centre, but, once there, will not be able to react as the original substrate would have done, and it will remain jammed in the active centre, stopping it from reacting with the genuine substrate. The process is exactly analogous to trying to open a door with a key that is just slightly wrong for the lock; we can put it in, but cannot take it out again, and the lock is jammed.

As both substrate and inhibitor can be regarded as fighting one another for the same active centre of the enzyme, this type of inhibition is called *competitive*. One interesting example is provided by certain bacteria, which need the substance para-aminobenzoic acid for growth. Traces of this substance, which occurs in blood and tissue, allow the bacteria to flourish. The closely related sub-

stance sulphanilamide, however, will compete with para-aminobenzoic acid for the bacterial enzymes, jamming them and hence
preventing bacterial growth.

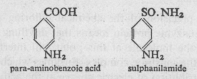

para-aminobenzoic acid   sulphanilamide

Hence the efficacy of the sulphanilamide drugs in preventing
certain bacterial-caused illnesses. Many other drugs and antibiotics have been discovered by using a similar principle, of finding
an enzyme essential for the bacteria but not to the host tissue and
then searching for a competitive inhibitor for it.

There is another class of inhibitor that has no affinity at all for
the enzymic active site; in fact the inhibitor can only bind to the
enzyme when it is complexed with its substrate. Thus this type of
inhibitor works by preventing the splitting of substrates into
products. Because there is no competition with the substrate for
the active site, this is called uncompetitive inhibition.

Multisubstrate and allosteric enzymes

We have said that the active site of most enzymes only involves a
small part of the protein molecule and that the major part of the
amino acid chain has very little involvement in the catalytic action.
But there is a certain group of enzymes where parts of the molecule
far removed from the active site are essential for the functioning
and control of enzyme activity. These enzymes were recognized
because when a V/S curve was plotted for them, the shape, unlike
that of Figure 11, was not hyperbolic but sigmoid (like a drawn-out
S). This shape can easily be explained if the enzyme has more than
one binding site for the substrate and the binding of the first
substrate molecule facilitates binding of the second or subsequent molecules, by altering the shape of the enzyme molecule,
a phenomenon known as co-operativity. The existence of enzymes
with such properties makes it much easier for the cell to respond
to a large demand for a particular product. Enzymes that have

more than one binding site are called allosteric (from the Greek *allos* – other, and *steric* – space).

Allostery, however, is not confined to multisubstrate enzymes. Some enzymes possess binding sites for substances other than the substrate and the effect of these substances may be to cause the enzyme molecule to change shape, thus inhibiting or facilitating substrate binding. Allosteric inhibition does not involve a substrate and an inhibitor vying for the same site, and is therefore called non-competitive inhibition (to distinguish it from the uncompetitive inhibition we have already mentioned). The exact mechanism of these allosteric effects is a fertile field of research for molecular enzymologists at the present time. But whatever their mechanism, it has become apparent that very many of the subtle controlling influences on enzyme activity that are necessary for the smooth running of the cell, which we discuss in Chapter 11, are mediated by such allosteric effects.

Enzyme classification

Whilst we will subsequently have much to do with enzymes in action and the reactions they catalyse, one further word ought now to be said about them. Despite the vast number of different enzymes and reactions that they catalyse, they can be classified as all conforming to a small number of general types. Before proceeding with the classification, though, we should begin by noting that enzymes are generally labelled with the suffix -ase, preceded by the name of the substrate on which the enzyme acts. Thus esterases hydrolyse esters, succinic dehydrogenase removes hydrogen from succinic acid, and so forth. There are some exceptions, though; pepsin and trypsin, for example, and one or two other old established enzymes have resisted most efforts to rationalize nomenclature. Several independent schemes existed for classifying the enzymes until the Commission on Enzymes of the International Union of Biochemistry brought a little logic to bear on the problem in 1964 by standardizing the nomenclature. They decided on what is perhaps the most logical way to subdivide the enzymes, i.e. to group them together according to the type of reaction they catalyse. This gives each enzyme its systemic name, but within these major groups each enzyme is identified by means of a classification number plus a name for everyday use.

The six major subgroups are usually descriptive of the enzymes they contain:

Group 1: *Oxido-reductases*, of which alcohol dehydrogenase is a prime example, catalysing the reaction

$$C_2H_5OH \rightarrow CH_3CHO + H_2$$

alcohol   acetaldehyde

Group 2: *Transferases*, which catalyse reactions of the type

$$A—B + C \rightleftarrows A + B—C$$

in which B is switched from A to C. A typical case is the synthesis of polysaccharides when the enzyme maltase can start with two maltose molecules, each containing two glucose units joined head to tail, and convert them into triose and glucose:

glucose—glucose + glucose—glucose $\rightleftarrows$ glucose—glucose—glucose
                                                                 + glucose

(maltose)   (maltose)   (triose)

thus adding one extra unit to the growing polysaccharide chain.

Group 3: *Hydrolases*, which split an intramolecular bond by the addition of H_2O. They are typified by the digestive enzymes and several others such as the esterases which we have already mentioned.

Group 4: *Lyases*, which catalyse bond-breaking reactions other than those of hydrolysis, for example the removal of a group attached to a quadrivalent carbon atom to leave a double bond.

Group 5: *Isomerases*, which catalyse the molecular rearrangement of their substrates; with phosphoglucomutase, for example, the enzyme lifts the phosphate off one part of the glucose phosphate molecule and fixes it to another part of the same molecule.

Group 6: *Ligases*, which catalyse bond-forming reactions, or the synthesis of a larger molecule by linking two precursor molecules.

Examples of all these types of enzyme in action will be found in the next few chapters.

CHAPTER 6

PATHWAYS OF METABOLISM

The reactions catalysed by purified enzymes can be studied at leisure and with comparative ease. Most enzymes when purified are stable for quite long periods provided they are kept cold, and a biochemist who wishes to study the kinetics of an enzymic reaction can go to the laboratory in the morning, take a bottle of enzyme crystals or solution out of the refrigerator, mix it with substrate, and analyse the reaction mixture at various times to see how much product has been formed. But, in the living cell, conditions are not so simple. There are many enzymes present, and the product of one enzymic reaction is itself likely to be the substrate of a second. Reactions occur very rapidly and within a very few seconds or minutes the molecules of a substrate may have been passed through the hands of a dozen or more enzymes, and have been transformed out of all recognition from the original starting substance. Molecules of glucose entering the cell, for example, are rapidly converted into an entire range of products. Some glucose is made into amino acids, some broken down to fatty acids and then converted into lipids, some oxidized to carbon dioxide. Every one of these interconversions is carried out by a series of enzymes acting in sequence, and the whole procedure is as ordered and efficient as that of the mass production line in which sheet steel enters at one end and cars or washing machines roll off the other.

On the mass production line, in which each operation depends for its success on those that have gone before, so that a slight alteration in procedure or mistake at the beginning of the line affects everything that comes after, any interference in the operation of the line produces not a car or washing machine but a mass of mangled metal instead. Thus the biochemist, trying to follow and interpret the cells' production lines, all too frequently finds that in order to do so it is necessary to throw so large a spanner in the works that everything grinds to a halt. But a knowledge of the route travelled by substrate molecules from beginning to end of their transformation by the cell is of great importance if we are ever to get a full picture of the way the cell works. So the effort must be made to draw up the 'flow sheets' for the chemical interconversions that the cell achieves.

INHIBITORS

Of the many ingenious methods that have been used to follow these interconversions, or metabolic pathways, as they are called, we may cite two here which have, time and again, proved their utility. The first demands the selection of a substance which specifically blocks one of a series of enzymes acting in sequence. Consider the reaction sequence

$$K \rightarrow L \rightarrow M \rightarrow N$$

where K, the added substrate, is converted by enzyme K-ase into L, which in turn is changed by L-ase into M and then by M-ase into N. What will now happen if we add an inhibitor which prevents the conversion of M into N by blocking M-ase? M will go on being made, but will not now be further utilized. So the amount of M present in the system will steadily build up. As its concentration increases, a time will come when the equilibrium of the reaction $L \rightarrow M$ begins to tilt in favour of L. When this happens the production of M from L will slow down and finally cease completely. But meanwhile enzyme K-ase will still go on converting K to L and hence the amount of L in the system will also begin to increase. Finally, K-ase will grind to a halt.

What is the effect of this? If we were to analyse the tissue in the absence of the inhibitor of M-ase, we should only have found K and N present, for L and M were used as fast as they were made. But with the inhibitor present, analysis will reveal the presence of L and M as well. So we know that L and M must be formed en route between K and N. It is as if we had wanted to study the make and type of cars and lorries going past on a fast road. Under normal conditions, they go past too quickly for us to be able to spot the make. But if we put in a traffic light on the road, the vehicles slow down and stop, and finally a queue of stationary cars builds up which we can examine at leisure.

Where can such inhibitors be obtained? In some cases it is

possible to use specific molecules that, because of their similarity to the natural substrates, act as competitive inhibitors, and much chemical ingenuity has gone into formulating them. In other cases it is possible to use the recently developed techniques of immunology to prepare specific antibodies which bind to and interfere with the biological functioning of particular proteins; such inhibitors are amongst the most specific of all.

Study of the sequence K → N using such inhibitors shows that L and M are intermediates. We still do not know the order of the reaction. Either

$$K \rightarrow L \rightarrow M \rightarrow N$$

or

$$K \rightarrow M \rightarrow L \rightarrow N$$

might be possible. A combination of things must help us decide the right sequence. First, we may be able to isolate the enzymes K-ase and L-ase and show that K-ase produces L and L-ase makes M. But other enzymes might also exist; we don't know. Then we may show that, in the presence of the inhibitor, M accumulates *before* L, indicating that M utilization is the first to be blocked, and L utilization only secondarily. Finally, we may use our knowledge of the chemical structures of L and M. We know that each enzyme catalyses only one reaction, and hence we may expect to find that chemically L is more like K, M more like N. It is like the party game in which we have to transform one word into another by changing one letter at a time, with the proviso that all the intermediate words also make sense. Suppose we have to get from BOAT to LEND and we know that our path must go via BENT, we can get there by way of the reaction pathway

BOAT → BEAT → BENT → BEND → LEND

and no other arrangement of the three intermediate words will fit the rules. Each enzyme can only change one letter in the word, and, if we read the rules of the game aright, it should be possible to put them in the right order. Examples of the use of inhibitors to study reaction pathways of this sort are common, and most of the metabolic pathways we shall later discuss (Chapter 8) have been studied by these methods.

ISOTOPES

The second method universally employed to follow metabolic pathways is that of radioisotopes. Normal carbon has an atomic weight of 12, but there also exists an *isotope* of carbon with an atomic weight of 14. Like all isotopes, it is chemically absolutely identical to the 'normal' atom and takes part in all its chemical reactions. The one difference is that carbon 14 is radioactive whilst carbon 12 is not. Molecules of carbon compounds containing the radioactive isotope in place of the 'normal' one can now readily be made. If we replace the carbon atoms of glucose by those of carbon 14, the glucose will be radioactive. If this glucose is now given to the cell to metabolize, all the substances into which the glucose is converted will begin to accumulate a proportion of the 'labelled' carbon 14 atoms. Any substance that lies on the metabolic pathway from glucose will thus become radioactive. By analysing the tissue and identifying the radioactive compounds, we will be able to show into which substances glucose is, and into which it is not, converted. What is more, if we perform such analyses at various times we will discover the *order* in which they lie on the pathway. Take the reaction sequence

$$A-X^0 \rightarrow B-X^0 \rightarrow C-X^0 \rightarrow D-X^0$$

in which we start with a small amount of compound A labelled with radioactive X^0. As soon as the reaction starts up, radioactive $B-X^0$ will begin to be formed and the amount of radioactivity in $B-X$ steadily increase. As more and more $B-X^0$ is formed, $C-X^0$ will begin to get radioactive also, and then $D-X^0$. Meanwhile, as $A-X^0$ becomes used up, the amount of radioactivity in $B-X^0$ will slowly decline. Finally, all the radioactivity will be found in $D-X^0$. We can plot a graph of the amounts of radioactivity in the different intermediates at various times (Figure 15). A study of this graph will immediately reveal the reaction sequence for which we are looking.

Radioactive isotopes used in this way are called 'tracers'. Apart from carbon, the radioactive isotopes of phosphorus and sulphur, and sometimes also the isotopes of nitrogen, oxygen, and hydrogen, are used in metabolic studies.

A more complicated problem arises where one particular compound is used as a precursor in the manufacture of many different substances. If the precursor is uniformly labelled, for instance if all its carbon atom are in the ^{14}C isotope form, many intermediates will subsequently be found to contain some of the radioactivity and it is very difficult to tell which particular pathways these intermediates are on. However, it is also possible to specifically label one or two particular atoms in a precursor molecule, and as different parts of the degraded molecule proceed down different pathways, it is easy to see, using these particular tracers, exactly how the precursor is utilized. The successes of the tracer methods have been enormous. For Rudolph Schoenheimer in the 1940s they revealed for the first time the 'dynamic state of body constituents' with which the whole of metabolic biochemistry is now concerned; to Melvin Calvin in the 1960s they gave the clue to the initial stages of carbon dioxide fixation in photosynthesis, and even more recently they have been used to study the mechanism of the incorporation of 'labelled' amino acids into protein during protein synthesis. Without them, a large part of our knowledge of metabolic activity would be impossible.

FIGURE 15. *Metabolic pathways shown by isotopes*

Radioactivity beginning in A—X^0 is slowly transferred to D—X^0 through B—X^0 and C—X^0, revealing that reaction sequence is A—X → B—X → C—X → D—X.

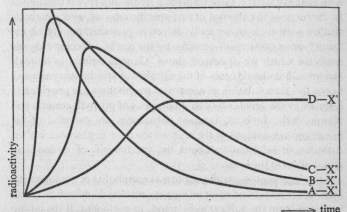

HOW A METABOLIC EXPERIMENT IS MADE

The simplest forms of metabolic experiments use whole animals. If an animal is fed on food of known composition, or containing radioactive tracers, and changes in the contents of the bloodstream, urine, and faeces are analysed, some indication of the fate and utilization of the food by the body can be obtained. But the potential of such a method is clearly limited; the information we can obtain is confined to the first and last stages of metabolism, and even the weight that we can put upon the results we do get is not certain. The conclusions we can draw are on the same level as those to be made about the activities and behaviour of a house full of people based on their grocery orders and the contents of their dustbins; revealing, but quite likely also misleading. But we can make such statements as 'when protein is fed to an animal what nitrogen does not appear as faeces is largely recovered in the urine', indicating that the animal is in 'nitrogen balance', and neither gaining nor losing nitrogen.

The intact animal can be 'improved' for experimental purposes if it is rendered abnormal in some way, by genetic malfunction, by illness, or by operation. Genetic defects, or mutations, are used widely in the study of bacterial metabolism, where they can be readily induced, for example through irradiation by X-rays or from a radioactive source. Genetic defects frequently reveal themselves in the form of the absence of one specific enzyme, and metabolic studies with such enzymically defective preparations are of the same type as those made possible by the use of a specific enzymic inhibitor which we discussed above. Genetic defects in animals are rarer, but classic cases of the absence of specific enzymes and hence the accumulation of abnormal metabolites are provided in humans by the genetically carried diseases of phenylketonuria and alkaptonuria. In both, unusual substances are excreted in the urine, and the analysis of the reasons for their appearance has led to valuable information about the mechanism of amino acid metabolism in the body.

Thus in alkaptonuria the abnormal metabolite is homogentisic acid, and it was found to be no longer excreted if amino acids were excluded from the sufferer's diet; and, in particular, if the amino acid tyrosine was not present. It could then be shown that homogentisic acid lay on the pathway of tyrosine breakdown, and thus several steps in this pathway could be revealed. But despite such successes, the use of whole animals to provide metabolic information is limited.

The next stage down from the use of the intact animal is to dissect out a whole organ – heart, or kidney, or brain – and to substitute for its normal blood supply an artificial system of arteries and veins carrying to and from the organ a substitute blood. Such preparations are called *perfused* organs, and can be kept alive, or functioning to all intents and purposes normally, for many hours, provided they are kept at body temperature and supplied in their artificial blood stream with oxygen and glucose. By comparing the composition of the perfusing fluid entering via the arteries and leaving via the veins, much information can be obtained about the uptake and utilization of various substances by the tissue.

The development of such perfused systems has been accelerated by the obvious value of such techniques to surgeons wishing to isolate temporarily some of the organs of the body whilst performing difficult operations. They are of course also a step on the way to the development of the 'artificial' kidneys and hearts which are now in common use in hospitals. The biochemist can learn a lot from them. Thus the comparison of perfused liver and brain shows that there are considerable differences in the ability of these organs to take up and use glucose and amino acids such as glutamic acid. Recognition of these differences led to the finding that brain shows a tendency to convert added glucose into glutamic acid, as compared to liver, which either oxidizes its glucose or uses it to synthesize glycogen. Such differences provide revealing glimpses into the way different organs are biochemically specialized to carry out different physiological functions, but they still fail to supply some of the crucial information about, for instance, the exact pathways by which glucose is converted into glutamic acid or glycogen.

It is only at a simpler level of organization than the whole organ that the biochemist finds a system which can be freely manipulated and controlled, whilst still retaining enough of the characteristics of the organ of origin to be fairly sure that it is revealing a 'real' and not a misleading metabolic route-map. This new and most useful tool was provided by the observation that, if thin slices of tissue are prepared and rapidly transferred into a warm, oxygenated medium of suitable composition, the slices will continue for several hours to take up oxygen and give off carbon dioxide, at a rate not very different from that of the intact organ. This survival depends partly on the preparation of the slices, partly on the composition of the medium. With the medium, one is providing the tissue slice with an artificial environment that must contain many of the conditions of the real one if the slices are to survive at all. The most essential things about it are that it must contain a buffer to keep the pH at or about 7.4, small amounts of sodium, potassium, calcium, and magnesium ions, a source of substrate for the slice, such as glucose, and a steady supply of oxygen. No other *organic* substances are required at all; the slice will continue to behave in this simple medium as if it were still part of the organ from which it was cut and were still being provided with a regular supply of blood by way of its capillaries. The only other requirements are that the slice is made rapidly on removal of the organ from the body, and that it is cut sufficiently thinly to make sure that the oxygen dissolved in the medium can get freely to all parts of it and enable its cells to breathe. In practice, the slices can be made about 0·3 and 0·5 millimetres thick and 1 centimetre square, by cutting free-hand with a razor-blade. The beauty of it is that the experimenter can precisely control all the variables of the experiment; change the composition of the surrounding medium, feed the slice any one of innumerable possible substrates alone or in combination, start and stop the reaction at accurately determined times, and measure precisely how much oxygen has been used up, carbon dioxide produced, and products accumulated.

It is hardly surprising then that for many years now the study of the behaviour of such slices has continued to provide a goldmine of useful information. The detailed mechanisms of glucose breakdown, amino acid interconversions, and fatty acid synthesis have been resolved in slice preparations. And, more and more, it is becoming possible to use them to tackle much more general problems of the regulation and control of the whole pattern of cell metabolism by such hormonal regulators as insulin and thyroxine, and by such seemingly simple ions as potassium and calcium. Using the slices, it is possible to obtain a picture of the cell functioning as an integrated whole, with all its parts in harmonious interconnection. And it is this picture that is the theme of contemporary biochemistry.

At a level below the slice, techniques based on enzymic digestion of the tissue, or gentle mechanical disruption, can separate individual cells from one another, whilst leaving each cell relatively intact and undamaged. For organs composed of several cell types, each of which may contribute its share to the biochemistry of the total, such techniques, followed by the separation of the cells into their separate classes by density gradient centrifugation (see Chapter 4), are proving of increasing utility.

At the final level below the slice, the tissue may be completely disintegrated by homogenization in the laboratory equivalent of a kitchen mixing machine, or by grinding with a pestle and mortar.

As has already been mentioned in Chapter 4, if the homogenization is conducted in an appropriately buffered salt or sucrose solution, the subcellular constituents may be preserved intact and separated by centrifugation. Different metabolic sequences can then be localized to particular cell compartments.

More robust homogenization will destroy the subcellular organelles as well, but will release into solution many enzymes whose activity cannot otherwise be easily studied because it is for various reasons masked in the slice or intact organ. Kinetics of individual enzyme reactions are much more easily followed in the homogenate than in the slice or organ, and for this reason the homogenate may be used to study the more detailed mechanism of a reaction first observed at a higher level of tissue organization.

A 'complete' statement of the biochemistry of the cell must not only be able to list what enzymes are present in the system, but where they are located within the cell, whether they are maximally active or lying dormant for lack of substrate or cofactors, and how their activity relates to that of all the other cellular components. The reasons for this will become more apparent when we turn to consider in due course the control of metabolism (Chapter 11). For the moment we should emphasize that each of the methods described above provides partial answers to partial questions. At the same time, each is likely to produce artefactual results. Inhibitors may have side effects on other enzymes or there may be otherwise unused sidepaths which become activated. The use of tissue preparations like slices may alter factors like ionic environment or oxygen supply, or enable vital metabolites or cofactors to leak into the surrounding medium. Cell and subcellular preparations may produce varying degrees of damage to the particles whose properties are being studied.

To examine the cell and its biochemistry it is necessary to interfere in some way with its normal functioning, and interpreting the results of this interference in terms appropriate to the 'reality' of the cell *in situ*, rather than the experimentally reduced artefact, requires the highest degree of critical awareness on the part of the experimenter. One must never forget that the substrate, enzyme, or cell is never really an isolate. Our experiments have to make it so to study it, but to interpret them, we must consider always the whole system. Metabolic biochemistry is fun, but it has its dangers!

From these general considerations we can now turn immediately to a description of some of the more important of the metabolic pathways that these methods have revealed.

CHAPTER 7

ENERGY-PROVIDING REACTIONS

There is more than one way of classifying metabolic pathways. In the older text-books, the division is normally made between pathways of the breakdown of already existing substances to simpler compounds – for example, the degradation of protein to amino acids and the subsequent breakdown and oxidation of the amino acids – and the reverse process, that of the synthesis of complex from simple substances – the synthesis of protein from amino acids. The breaking-down reactions are called *catabolic*, the synthetic reactions are *anabolic*, and the sum of all the catabolic and anabolic reactions occurring at any time represent the total of the cell's *metabolism*. Another way of looking at the same reactions though is to look at the standard free energy change that occurs during their course, to divide them into exergonic or endergonic (energy yielding or energy using) reactions. However, any energy change that occurs in the cell overall has of course to be negative. This is why the cell never carries out the two types of reaction in isolation from each other; exergonic ones are always linked to endergonic reactions, and the one is used to drive the other. In any one pathway the two types of reactions are linked by a common chemical intermediate, which possesses a highly reactive group. The intermediate with its reactive group is formed at the expense of one of a small number of compounds, ubiquitously present in the cell.

There are three main types of such *group transfer molecule*; the amidine phosphates like creatine phosphate; the thioesters like Coenzyme A, which we will come across many times in subsequent discussions; and the nucleotide phosphates. The most important of all of the group transfer molecules, and one of the most universally important biochemical molecules, is adenosine triphosphate, ATP (page 41). ATP is a highly reactive substance which can readily donate its terminal phosphate group to other molecules, thus phosphorylating them. It can also act as an adenylating agent, in which case the whole of the adenine-ribose

127

part of the molecule is transferred to another compound (p. 137). The concept of energy coupling by group transfer molecules via common intermediates is so crucial to an understanding of how the metabolic pathways within the cell are ordered, that we must look first at the way ATP is generated and leave consideration of the way in which this ATP is used until a subsequent chapter.

The cell gets most of its energy by oxidizing foodstuffs. Oxygen is taken up from the air by way of the lungs, and enters the bloodstream to form a loose combination with the red, iron-containing blood pigment, haemoglobin. When the circulating blood reaches the body tissues, where the amount of oxygen is low, the haemoglobin-oxygen complex dissociates, releasing oxygen into solution, whence it can diffuse through the capillary walls and into the cells. In return, carbon dioxide produced by the cells passes into the bloodstream and is swept away to the lungs, where it is released into the air. Thus the cell is kept constantly supplied with enough oxygen to help produce the energy it needs.

Oxidations are reactions which tend to have a very high free-energy yield; we have already quoted the figure of 2,820 kJ released during the oxidation of one gram-molecular weight of glucose. But so large an amount of energy must not be released all at once; instead, its release is the result of the sequential operation of a series of enzymes; together, they form the first of the metabolic pathways that we shall discuss in detail.

Let us first consider the nature of the oxidation reaction itself. Asked to write its equation, we would normally produce

$$A + O_2 \rightarrow AO_2 \qquad (1)$$

where A, an oxidizable substrate, is *oxygenated* to AO_2. But, whilst the adding of oxygen to a molecule is *one* form of oxidation, there are others. One such is the *removal of hydrogen*, according to the reaction

$$AH_2 + X \rightarrow A + XH_2 \qquad (2)$$

In reaction (1) we oxidized A at the expense of oxygen. In reaction (2) we have removed hydrogen, i.e. oxidized A, at the expense of another substance, X, which has been correspondingly reduced. In reaction (2) an oxidation has been performed in a reaction into which oxygen itself does not even enter.

There is also a third form of oxidation. In order to follow this, we must remember that the atom of hydrogen can be regarded as composed of the charged hydrogen ion, H^+, plus one *negatively* charged electron, e^-. In aqueous solutions, where the ions H^+ and OH^- both exist, *the removal of an electron from a charged ion is, in many respects, equivalent to its oxidation*. Thus when the doubly charged ion of iron, the *ferrous* ion, is converted to the triply positive, *ferric* form, an oxidation reaction has also occurred:

$$Fe^{++} \rightarrow Fe^{+++} + e^- \tag{3}$$

All these reactions occur in the cell, but on the whole, the cell finds the second and third easier to perform, and, what is more, can perform either of them without the direct intervention of oxygen itself. Reaction (1) does have some importance though. There are enzymes, called oxidases, which catalyse the direct oxidation of their substrates by atmospheric oxygen. They are found in most cells, but are especially common in plants. The effects of one of them, polyphenol oxidase, are familiar to those who peel their apples before eating them. Apples contain traces of a compound called catechol which is oxidized, under the influence of polyphenol oxidase, to a complex, dark brown substance. When an apple is cut its surface is exposed to the air and polyphenol oxidase can begin to work, turning the surface of the apple gradually brown.

But in general the cell prefers to use equation (2). In principle, the cell's oxidative apparatus consists of a series of substances like the X of that equation and also of the enzymes to catalyse the reactions. Substances like X we class as *hydrogen carriers* or, for reasons that will become clear shortly, *electron carriers*. The enzymes which catalyse the reactions are *dehydrogenases*. They are usually specific for the individual substrates that they oxidize, so there are dehydrogenases for every substance the cell can oxidize. We have already met lactic dehydrogenase, for instance (see page 104). On the other hand, the hydrogen carriers themselves are not all specific; the same set of carriers serves for practically all the oxidations performed in the cell, once the specific dehydrogenase has made the initial link-up by transferring hydrogen from A to X in the equation.

The hydrogen carriers are arranged in a strict linear sequence which is maintained because they are all firmly embedded in the lipoprotein of the inner membrane of the mitochondrion within the cell. As this is also the site where food substances are oxidized the whole process of hydrogen transfer from food to the terminal acceptor oxygen with the subsequent release of energy can take place quickly and efficiently. The carriers are mainly proteins carrying prosthetic groups in the form of metallic ions that are capable of being alternately oxidized and reduced. The measure of this oxidation/reduction capacity is termed their redox potential. Electrons are always transferred from substances with a high redox potential to those of lower potential. So this in part helps to explain the requirement for linear ordering of the carriers.

The hydrogen transport process is started with a high potential energy substrate, AH_2, whose hydrogen is passed down a chain of carriers, each time releasing a little energy, which can be utilized for the synthesis of a group transfer molecule such as ATP, until at the very end the last carrier transfers its hydrogen to oxygen and thus disposes of it as a molecule of water. The enzyme for this last reaction is thus an oxidase:

$$ZH_2 \; + \; (O) \; \rightarrow \; Z \; + \; H_2O \tag{4}$$

The process works like one of the water-wheels that used to be used for grinding corn. Water pours into the top bucket of the water-wheel at high potential energy, as it has several feet to fall. The first bucket breaks its fall and in doing so itself begins to move, carrying the wheel round. The water spills into the second bucket, yielding more energy and hence more movement, then into the third, the fourth, and finally into the stream at the bottom; during its fall down the series of buckets its potential energy has been tapped off at several points and used to spin the wheel. As with the water, so the hydrogen in its fall towards oxygen is checked at several points and its available energy tapped off.

NAD

The identity of most of the hydrogen carriers is now well established. They are arranged in three groups, the redox potentials of the carriers within each group being roughly similar, while each group is linked by carriers with potentials mid-way between the value of the potentials of the groups they are linking. When the dehydrogenase enzymes were first studied, it was found that, if they were dialysed, they lost activity. A soluble cofactor was obviously leaking away, leaving the protein enzyme in the dialysis bag inactive. The cofactor was isolated and christened *coenzyme I*, but was subsequently identified as substance X of our equations, the first of the hydrogen carriers. Obviously, in its absence, reaction (2) cannot proceed; the dehydrogenase has nothing to do with the hydrogens that it collects from A. Chemically, X, or coenzyme I, turns out to be a nucleotide substance: nicotinamide adenine dinucleotide, or NAD for short. We can write it to show its composition more clearly as:

nicotinamide-ribose-phosphate-phosphate-ribose-adenine

The business end of the molecule, the part that actually accepts the hydrogen, is the nicotinamide nucleotide, which reacts like this:

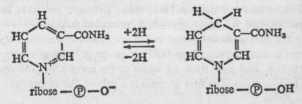

Nearly all the dehydrogenases pass hydrogen atoms on from their substrates direct to NAD. The hydrogen atoms arrive at NAD from many different points, like commuters descending upon a railway station, but, once having been taken on board, they all run the same common pathway to molecular oxygen at the terminus of the line. There is one branch line which enables the hydrogens from some substrates to bypass NAD completely, and there is also at least one other mainline to oxygen, which starts from a substance closely related to NAD – nicotinamide adenine dinucleotide phosphate, or NADP – the significance of which we shall discuss in the next chapter. But these alternative routes are of minor significance compared with the NAD-line.

Linked with NAD are a group of iron and sulphur containing proteins, the ferredoxins, which form a sort of marshalling yard.

If the next station along the line from NAD is temporarily full up, the hydrogen is passed between NAD and the ferredoxins until it can move on. Following NAD there is a series of stations en route for oxygen – as many as twenty have been postulated at one time or another, although there are probably only seven or eight on the main line route.

All the H atoms arriving at NAD travel on through each of these stations in turn before arriving at the oxygen terminus. They do so very quickly – the time of transit down the hydrogen carrier line is in the order of thousandths of a second, and this rapidity, as well as the complexity of some of the chemical reactions involved and the fact that the reactions are coupled to the production of ATP, has made the study of the chemical nature of the hydrogen carriers very difficult. Despite the efforts of many biochemists over fifty years or more, including such people as Otto Warburg in Germany, David Keilin in England, and, more recently, David Green and Britton Chance in America, there are many mysteries still remaining. The scheme we present here, then, although probable on present evidence, is almost certainly by no means final.

The first two stations beyond NAD are substances not so very different to it in structure. Like NAD, they accept hydrogen atoms by *reduction*, and yield them up again to the carrier that follows them by oxidation. The first of them is a yellow, fluorescent substance, *flavin*.

Flavin

The flavin molecule consists of three linked rings:

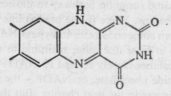

but the part of the molecule that actually does the work is the pair of nitrogens on the centre and right-hand rings, which can be alternately oxidized and reduced:

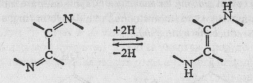

Like other biologically active substances we have met, flavin is found combined as a nucleotide, linked to ribose phosphate and sometimes also to adenine nucleotide, forming flavin mono-nucleotide (FMN) or flavin adenine dinucleotide (FAD). The flavin nucleotides, unlike NAD, are bound firmly on to proteins and cannot be removed by dialysis. Such proteins are known as *flavoproteins* (FP). The typical reaction in which the flavoproteins participate is the oxidation of $NADH_2$:

$$NADH_2 + FP \rightleftharpoons FPH_2 + NAD$$

In this reaction, the protein part of the flavoprotein may be regarded as catalysing the oxidation of NAD; it is thus an enzyme, *NAD dehydrogenase*.

One other FP enzyme is also known, that which oxidizes succinic to fumaric acid:

$$\begin{array}{c} CH_2COOH \\ | \\ CH_2COOH \end{array} + FP \rightleftharpoons \begin{array}{c} CH.COOH \\ \parallel \\ CH.COOH \end{array} + FPH_2$$

The existence of this enzyme, which means that hydrogen atoms from succinic acid can be passed directly on to flavin without travelling via NAD, represents the one major branch line so far found by which hydrogen atoms can be dispatched to oxygen but bypass NAD. It is not clear *why* the oxidation of succinic acid should proceed by a different pathway than that of most other substrates, but it is certain that the enzyme succinic dehydrogenase which catalyses the reaction is a fairly complex one.

Quinones

The next station beyond FP in the hydrogen carrier line (although one in the same redox potential group) is almost certainly yet another alternately oxidized and reduced molecule, this time a substance called *quinone*, or *coenzyme Q*. Quinones are molecules that contain two oxygen atoms bound to a 6-carbon ring, and can readily be reduced to *hydroquinone*.

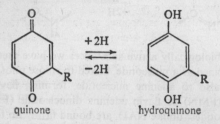

R may be one of a variety of residues; with only a little modification the molecule becomes that of Vitamin E. It is also closely related to Vitamin K. The hydrogen carrier, though, appears to be neither of these; R is a long chain fatty acid residue, and the substance is called coenzyme Q. Probably, coenzyme Q participates in the oxidation of reduced flavoprotein by accepting two hydrogens from it, and is subsequently oxidized in its turn by passing its hydrogens on to the next of the carriers, *cytochrome b*.

The cytochromes

With the next station along the hydrogen carrier line, we move from a type of hydrogen transport in which hydrogen is moved directly by the alternate addition and subtraction of hydrogen atoms as in equation (2) of page 128, to a region of the carrier line in which the transfer is represented by alternate oxidation and reduction of an ion of iron, Fe, as in equation (3).

The hydrogen atom splits into a hydrogen ion and an electron:

$$H \rightarrow H^+ + e^-$$

and the electron is carried down the following few stations of the hydrogen carrier line by means of the alternate oxidation and reduction of iron ions:

$$Fe^{++} \rightleftarrows Fe^{+++} + e^-$$

whilst the H^+ part of the hydrogen atom is temporarily 'parked' in the water surrounding the hydrogen carrier proteins until, at the end, it can be dragged back to react with oxygen.

The substances which form the stations on this part of the line between NAD and oxygen are a series of very closely related proteins called *cytochromes*. Each cytochrome contains, firmly bound to the protein surface, a prosthetic group of the iron-containing substance *haem*. The cytochromes are thus similar to the haemoglobin of blood, and, like haemoglobin, they are red in colour and can react directly with oxygen.

Hydrogen atoms from quinone are passed to the first of the cytochromes, *cytochrome b*, which traps the electron, using it to turn its ferric ion Fe^{+++} into ferrous ion Fe^{++}, and parks the H^+ in the surrounding medium. The electron is passed from cytochrome b to the second and third cytochromes, c_1 and c. The ferric ion of c is reduced to ferrous, and the ferrous ion of b is oxidized to ferric once more. In the same way, the electron is passed from cytochrome c to cytochrome a and then to cytochrome a_3. In fact, the cytochromes are working like a line of basketball players, passing the electron ball from one to the other, and each time getting closer to the goal of oxygen.

With cytochrome a_3, the goal is reached at last. This cytochrome differs from the others in that as well as possessing the iron haem group it is also closely associated with the copper ion Cu^{2+} and copper, like iron, can exist in the cuprous Cu^{2+} and cupric Cu^{3+} forms; the transfer of electrons to molecular oxygen probably occurs from the copper iron after it has been reduced by the cytochrome a_3 Fe^{2+}. So the final reaction in the electron transport chain involves two electrons, two hydrogen ions and an oxygen atom reacting together to form water:

$$2\,Cu^{++} \;+\; 2H^+ \;+\; O \;\rightarrow\; 2\,Cu^{+++} \;+\; H_2O$$

Because the cytochrome a_3 complex works by oxidizing the earlier cytochromes at the expense of oxygen, it is sometimes referred to as if it were an enzyme, *cytochrome oxidase*.

Incidentally, the oxidation of reduced cytochrome is prevented by cyanide, and it is this that accounts for the poisonous effects of cyanide in the body. If cytochromes cannot be oxidized, hydrogen transport cannot occur and ATP cannot be made. The result is a quick death through energy-lack. Some narcotic drugs, on the other hand, work by slowing down the rate at which *oxidized* cytochromes can be reduced by accepting electrons from further back in the hydrogen transport line, resulting in a small but significant drop in energy levels.

We can now draw in all the stations of the hydrogen transport line, in a sequence running from substrate to oxygen, including also the branch line from succinate which bypasses NAD:

$$succinate \rightarrow FP$$
$$\searrow$$
$$Q \rightarrow cyt\ b \rightarrow cyt\ c_1 \rightarrow cyt\ c \rightarrow$$
$$\nearrow \qquad\qquad\qquad\qquad cyt\ a \rightarrow cyt\ a_3 \rightarrow O_2$$
$$AH_2 \rightarrow NAD \rightarrow FP$$

We do not know whether, in drawing the chain like this, any essential sections have been omitted, nor can we be quite sure of the details of just how the branch line from succinic acid couples with the main NAD-line, but we may be certain that the main outlines of the map we have drawn are correct.

OXIDATIVE PHOSPHORYLATION

The point of sending the hydrogen atoms down a series of carriers on their way to oxygen is to tap off the available energy, so we may expect to find that their progress down the line is coupled with the production of ATP. Such indeed is the case, and, for oxidation to take place via the carrier line, it is necessary for ADP and phosphate to be present. During oxidation these substances are steadily used up, and ATP formed. When all the ADP or phosphate is finished, oxidation comes to a halt. The coupling mechanism is thus rather like the linkage between the engine of a car and its wheels, by way of the gears. Oxidation causes the engine to turn, and this in turn spins the wheels and forms ATP. But, if we stop the wheels from turning by putting on the brake or removing the ADP supply, the engine stalls and oxidation ceases. In a car, we can prevent this stalling by uncoupling the engine from the wheels and letting it tick over freely by putting the gears in neutral. The cell too has such a mechanism for uncoupling oxidation from phosphorylation, which we shall discuss later when we come to consider how the cell regulates its activities. But normally the two are firmly linked.

Three questions spring to mind at this stage:

(1) Why ADP?

(2) Which are the coupling points along the electron transport chain for ATP production?

(3) What is the mechanism of coupling?

The first two questions are relatively simple to answer. Adenosine, it will be remembered, is the nucleotide formed from adenine and ribose, and it can be phosphorylated to yield first adenosine monophosphate (AMP), then adenosine diphosphate (ADP) and finally ATP (see page 41). If we follow the course of its hydrolysis back to adenosine we find the following series of reactions:

$$\text{ATP} \xrightarrow{H_2O} \text{ADP} \ + \ \text{\textcircled{P}} \qquad \Delta G° = -31 \text{ kJ}$$

$$\text{ADP} \xrightarrow{H_2O} \text{AMP} \ + \ \text{\textcircled{P}} \qquad \Delta G° = -31 \text{ kJ}$$

$$\text{AMP} \xrightarrow{H_2O} \text{A} \ + \ \text{\textcircled{P}} \qquad \Delta G° = -12\cdot5 \text{ kJ}$$

Thus, while the first two reactions are highly exergonic, the hydrolysis of the third and last phosphate is much less so. In this, it resembles the majority of phosphate esters. It is the di- and triphosphate bonds of ATP which are exceptional. The important thing though is that all the reactions are fully reversible; the synthesis of ATP is thus highly endergonic and provides a convenient method of using the sometimes vast amounts of energy released from glucose oxidation. During the sequence of oxidation steps leading from glucose to oxygen, whenever a point is reached where a reaction occurs with a $\Delta G°$ of more than -31 kJ, this reaction can immediately be coupled with the synthesis of a molecule of ATP from ADP and phosphate. The ATP thus formed can then be utilized, at perhaps a different time and place within the cell, in the way we already mentioned at the beginning of this chapter.

If a quantitative study is made of the amount of ADP and phosphate used and ATP made per pair of hydrogen atoms passed down the line, it is found that three molecules of ADP disappear and three of ATP are formed during the transport of hydrogen from $NADH_2$ to oxygen. A clue as to the coupling points is provided by the observation that the number of ATP molecules formed during the oxidation of succinic acid is not three but two. As the succinic acid and NAD lines merge at the quinone stage, and are identical from then on, it follows that the *extra* ATP made on the NAD line must be formed *between* NAD and quinone, whilst two molecules of ATP are made between quinone and oxygen. More refined analytical techniques involving the isolation of submitochondrial complexes carrying different parts of the carrier chain, and the sophisticated use of substances which inhibit particular carriers, and artificial electron donors that can only feed into the chain at specified points have pinpointed the other two sites as lying between cytochrome b and cytochrome c, and between cytochrome a and O_2.

Unfortunately the answer to our third question, that of the mechanism of coupling, is the subject of heated controversy between various research workers and undoubtedly the exact nature of the gears that couple ATP formation to electron transport are still to be worked out. However, any hypothesis has to account for a number of observed experimental results, the most important of which are:

(1) Coupled ATP synthesis occurs only on intact mitochondrial membranes which surround a completely enclosed space.

(2) The enzyme responsible for ATP synthesis is a huge multi-subunit protein molecule and is firmly bound to the inner mitochondrion membrane. It can also in certain circumstances act in reverse, to hydrolyse ATP to ADP and (P), that is, as an ATP-ase.

(3) The energy released during electron transport is not always immediately used for ATP synthesis. It can somehow be stored to be used later, even after transport has stopped altogether.

(4) During the electron transport process, movement of protons (that is, H^+) across the mitochondrial membrane has been shown to occur.

A mechanism that incorporates these observations is the chemiosmotic hypothesis first suggested by Peter Mitchell in 1961. His ideas were greeted with much scepticism when they were first formulated but they have gradually come to be accepted by the majority of biochemists, so much so that Peter Mitchell has been awarded the 1978 Nobel Prize for Chemistry for this work and other contributions in the field of bioenergetics. It postulates that the electron carriers are arranged so precisely within the inner mitochondrial membrane that transfer of hydrogen ions, that is, protons, is always spatially in the same direction, from the inner mitochondrial matrix to the space between the inner and outer membranes. The result of this transfer is to establish a gradient of hydrogen ion concentration across the membrane. The hydrogen ions will tend to flow back down this concentration gradient across the membrane, and it is during this flow back that they can interact with ADP and (P) by way of the ATP synthesizing enzymes. The synthetic reaction for ATP takes place in the strongly hydrophobic lipid core of the membrane, where water is excluded and hence the reverse reaction of hydrolysis is blocked. Whilst this proposed mechanism for ATP synthesis during oxidation, that is, oxidative phosphorylation, does not command universal assent, it is the nearest one can yet come to a description of this fundamental and well-nigh universal process amongst living oxidative systems.

The synthesis of all the other group transfer molecules we mentioned at the beginning of this chapter are, like that of ATP, highly endergonic. Creatine phosphate (CrP) has the structure

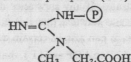

Like ATP, creatine phosphate can be hydrolysed to yield creatine (Cr) and inorganic phosphate (P), with a $\Delta G°$ of -31 kJ. It then becomes possible for ATP and creatine to transfer phosphate groups from one to another without either needing or releasing energy.

$$A—(P)—(P)—(P) \;+\; Cr \;\rightleftharpoons\; A—(P)—(P) \;+\; Cr—(P)$$

This reaction too has a part in the energy balance of the cell, for the creatine phosphate can serve as yet a further energy store; when all the available ADP has been converted into ATP, and the energy-bank is full, ATP begins to transfer phosphate to creatine to give creatine phosphate, releasing ADP once more to be converted again into ATP. On the other hand, when ATP levels are low, phosphate can be transferred back from creatine phosphate to ADP, replenishing ATP stocks. If the ATP-ADP system represents the current account of the cell's energy-bank, creatine phosphate is the deposit account.

We can thus describe the methods whereby the cell gets the energy it needs. Many different reactions can provide ATP. As many different ones subsequently draw on the energy locked up inside that molecule. Much of the rest of this book will be occupied by describing some of the ways by which ATP is made and used. We shall find ATP being made during glucose, fatty acid, and amino acid breakdown, and ATP being split in the synthesis of macromolecules, the contraction of muscle and the transmission of nerve, the transport of wanted substances into the cell and undesirable ones out of it; every phase of the cell's activity is dominated by the availability of this universal energy currency.

In its absence, the cell runs down, ceases to be able to perform its physiological functions, and in desperation begins to break open its last remaining food reserves. When ATP is plentiful, the cell relaxes, slows down the oxidation of glucose, and uses its superabundance of energy to build up long-term storage reserves of fats and carbohydrates. If any substance deserves to rank as central to biochemistry-as-kinetics, then it is ATP.

energy for synthesis and work

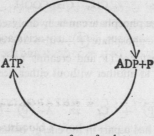

energy from food

It was for the recognition of the significance of this pivotal role of ATP that Fritz Lipmann, of the Rockefeller Institute in New York, was awarded the Nobel Prize for 1953.

CHAPTER 8

SOURCES OF ENERGY

FOOD

The substrates which the cell oxidizes in the manner we described in the last chapter are food. A unicellular organism, like amoeba, swallows its food whole – the cell literally engulfs particles of nutrient and sucks them into itself. In more complex animals, however, the food reaches the cell only after it has already been broken down into relatively small molecules which the cell can easily absorb and handle. This process of breaking down is called digestion, and its details are well known. In the human, food is broken down in the mouth, stomach, and intestine by a series of enzymes – amylases in the saliva and pancreatic juice for carbohydrates, pepsin in the stomach and trypsin and chymotrypsin in the intestine for proteins, and lipases in the intestine for fats. The resulting mixture of sugars, amino acids, and fatty acids is absorbed through the intestinal wall, and, directly in the case of the sugars and amino acids, indirectly in the case of fats, enters the blood stream. In the well-fed individual, the cell is thus kept steadily supplied with its food requirements as low-molecular-weight substances dissolved in the circulating blood from which it can draw at will. Under normal circumstances the bulk of the cell's energy requirements are met by the breakdown of sugars, although, as we shall see, both fatty acids and amino acids can also, if the need arises, be broken down and the energy they release used to synthesize ATP.

GLUCOSE METABOLISM

The glucose circulating in the body's bloodstream serves to keep the cells supplied with a steady source of potential energy. Many cells, notably liver and muscle, absorb from the blood more than they require for their immediate needs, and this surplus is turned into glycogen, to guard against harder times later on. If the glycogen stores are already full, the liver cells begin to convert the surplus glucose into fat, which can subsequently be deposited in storage depots in various convenient portions of the body. It is this primitive defence mechanism against the possibility of glucose-starvation that has results so unflattering to the vanity of those humans fortunate enough to be able to eat each day more glucose-providing substances than their bodies really need; but for other animals, and for the large proportion of the world's population that exists precariously on the borders of starvation, it is an absolutely essential mechanism for not wasting any potential supplies of energy. We shall consider the detailed mechanisms of the synthesis of fat and glycogen in the next chapter. For the moment our concern is with the fate of the glucose immediately utilized as an energy source.

The breakdown of glucose is accomplished in two stages. In the first, the 6-carbon glucose molecule is split into two 3-carbon fragments of pyruvic acid, releasing only a little energy in the process. This first half of glucose breakdown can be carried out in the absence of oxygen, and is called *glycolysis*. In the second stage, the pyruvic acid is completely oxidized to carbon dioxide and water by way of a series of acidic intermediates. This stage releases a great deal of energy, is oxygen-requiring, and is called *glucose oxidation*.

When a situation arises in which the cell needs a great deal of energy very quickly, it may happen that the demand for oxygen is so great that it outruns the supply available to the cell from the haemoglobin of the bloodstream. When this happens, the cell falls back on operating the glycolytic half of glucose breakdown without the oxidative half. The energy yield from this is smaller, but it does mean that the cell has a second line of defence should its oxygen supplies fail it. Glucose breakdown in the absence of oxygen is called *anaerobic* glycolysis. It results in the production of pyruvic acid which the cell reduces to lactic acid and then discharges into the bloodstream. Anaerobic glycolysis provides the energy for any sustained bout of violent muscular exercise – a hundred yards sprint, for example – and, indeed, after such exercise it is possible to measure large increases in the level of lactic acid in the athlete's bloodstream. The glycolytic pathway is also that followed in many bacteria and also in yeast when it ferments sugar to alcohol.

Glycolysis then is a practically universal pathway of glucose degradation, and has been studied for a long time in very many organisms. Much of the pioneer work was done with yeast, because of the exceptional interest that alcohol production has always had for scientists and non-scientists alike; it was only later that it was found that the same sequence of reactions which had been elucidated in yeast operated also in human muscle. The full discovery of the reaction sequence took many years, beginning with the Buchners in Germany and Harden and Young in England at the beginning of the century. The details were fully worked out by Embden and Meyerhof in Germany in the 1930s, and the glycolytic pathway is thus often known after them as the Embden-Meyerhof pathway. The oxidative part of the pathway, leading from pyruvic acid to carbon dioxide, was more difficult; it was not fully understood until the mid 1950s. Amongst the many who helped its unravelling, the names of Szent-Györgyi in Hungary, Krebs and Ogston in England, and Ochoa and Lipmann in America stand out.

As we might expect from what we already know about enzyme reactions, the cell goes a long and laborious way about the breakdown of glucose. No less than ten enzymes are required to take it down as far as pyruvic acid, and another ten are needed in the subsequent oxidation of pyruvic acid to carbon dioxide and water. The enormous importance of this reaction sequence makes it necessary for us to try to describe it in full here, particularly as it will also serve us as a model for many later reaction sequences which we shall mention but not consider in anything like such detail. For the reader in a hurry, a judicious turning over of the pages until Figure 16, page 150 and from there to Figure 17, page 153 would be permissible. But for the stronger-willed, here is the fuller story.

GLYCOLYSIS

The first step towards the breakdown of glucose is at first sight a strange one; it demands the *phosphorylation* of glucose. The fact is that in order for glucose to participate in the reactions that follow, it must be activated at the expense of ATP. Fortunately the ATP is later recovered, with dividends. The phosphorylation is carried out by the enzyme *hexokinase*:

$$
\begin{array}{l}
\text{CHO} \\
| \\
\text{CHOH} \\
| \\
\text{glucose} \quad \text{CHOH} + \text{ATP} \longrightarrow \text{CHOH} + \text{ADP} \\
| \\
\text{CHOH} \\
| \\
\text{CHOH} \\
| \\
\text{CH}_2\text{O} \boxed{}
\end{array}
\qquad
\begin{array}{l}
\text{CHO} \\
| \\
\text{CHOH} \\
| \\
\\
| \\
\text{CHOH} \\
| \\
\text{glucose-6-phosphate} \\
\text{CHOH} \\
| \\
\text{CH}_2\text{O}-\boxed{P}
\end{array}
\qquad (1)
$$

Glucose-6-phosphate is then converted by the enzyme *phospho-hexoisomerase* to fructose-6-phosphate. A further phosphate group is then transferred from ATP to produce the highly reactive molecule fructose 1, 6, diphosphate, catalysed by the enzyme phosphofructokinase.

$$
\begin{array}{l}
\text{C}\,\text{HO} \\
\| \\
\text{C}\,\text{HOH} \\
| \\
\text{glucose-6-phosphate} \quad \text{CHOH} \\
| \\
\text{CHOH} \\
| \\
\text{CHOH} \\
| \\
\text{CH}_2\text{O}-\text{P}
\end{array}
\rightleftharpoons
\begin{array}{l}
\text{C}\,\text{H}_2\text{OH} \\
\| \\
\text{C}\,\text{O} \\
| \\
\text{CHOH} \quad \text{fructose-6-phosphate} \\
| \\
\text{CHOH} \\
| \\
\text{CHOH} \\
| \\
\text{CH}_2\text{O}-\text{P}
\end{array}
\qquad (2)
$$

$$
\begin{array}{l}
\text{CH}_2\text{O} \boxed{} \\
| \\
\text{CO} \\
| \\
\text{fructose-6-} \quad \text{CHOH} + \text{ATP} \longrightarrow \\
\text{phosphate} \quad | \\
\text{CHOH} \\
| \\
\text{CHOH} \\
| \\
\text{CH}_2\text{O}-\text{P}
\end{array}
\qquad
\begin{array}{l}
\text{CH}_2\text{O}-\boxed{P} \\
| \\
\text{CO} \\
| \\
\text{CHOH} + \text{ADP} \\
| \\
\text{CHOH} \quad \text{fructose 1,6,} \\
| \qquad\qquad \text{diphosphate} \\
\text{CHOH} \\
| \\
\text{CH}_2\text{O}-\text{P}
\end{array}
\qquad (3)
$$

In the fourth reaction, the fructose 1, 6, disphosphate is split into two 3-carbon fragments by the enzyme *aldolase*:

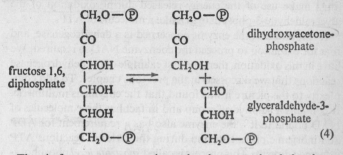

$$\text{fructose 1,6, diphosphate} \rightleftharpoons \text{dihydroxyacetone-phosphate} + \text{glyceraldehyde-3-phosphate} \qquad (4)$$

Thus in four separate reactions, the glucose molecule has been split down the middle into two separate substances, each containing three carbon atoms. At the same time, two molecules of ATP have been spent in phosphorylating the glucose, thus activating it and preparing it for breakdown. The stage is now set for these ATPs to be recovered. Dihydroxyacetone-phosphate and glyceraldehyde-3-phosphate are in fact isomers, having the same overall formula, though different structures. In all the further reactions of the sequence, only glyceraldehyde-3-phosphate is used by the cell. So another enzyme now converts the dihydroxyacetone-phosphate into the isomeric glyceraldehyde-3-phosphate, thus making sure that both halves of the original hexose molecule are used up. The enzyme concerned is *triose phosphate isomerase*:

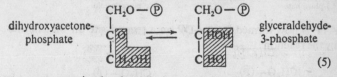

$$\text{dihydroxyacetone-phosphate} \rightleftharpoons \text{glyceraldehyde-3-phosphate} \qquad (5)$$

The next stage in the degradation is the *oxidation* of glyceraldehyde-3-phosphate to 3-phosphoglyceric acid. In principle, one could write the reaction as

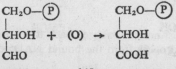

145

but in fact the details of the reactions are more complex.

Oxidation, we know, is an energy-yielding process, and the cell must make use of the energy released during oxidation of the glyceraldehyde-3-phosphate to make a molecule of ATP.

It is found that the enzyme concerned is a dehydrogenase, and that for oxidation to proceed the coenzyme NAD is required. We have in this oxidation, then, the first example of the dehydrogenase reactions that we discussed in the previous chapter. To add complexity to the picture it was found that the enzyme is multimeric. It contains a sulphydryl group and in fact has three molecules of NAD bound to it – the enzyme also has a requirement for ADP and inorganic phosphate and during the series of reactions ATP is formed directly. This process is called *substrate level* phosphorylation to distinguish it from ATP synthesis coupled to the electron transport chain, which is oxidative phosphorylation.

The reaction occurs in three stages. In the first stage, the sulphydryl group of the enzyme and the NAD are reduced and glyceraldehyde-3-phosphate is phosphorylated by inorganic phosphate to 1,3, diphosphoglyceric acid:

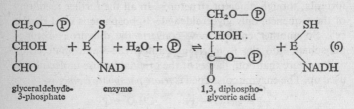

ATP is now formed from ADP on a second site on the enzyme by the reaction

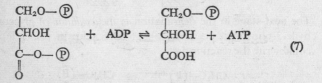

The enzyme must now be oxidized back to its original form, by transferring hydrogens from the bound NAD to NAD free in

solution in the cell's cytoplasm. Thus the third stage of the reaction is

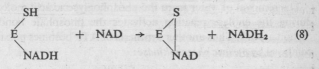

$$E \begin{array}{c} SH \\ | \\ NADH \end{array} + NAD \rightarrow E \begin{array}{c} S \\ | \\ NAD \end{array} + NADH_2 \qquad (8)$$

One molecule of glucose yielded two of glyceraldehyde-3-phosphate, at the expense of two of ATP. Now, two molecules of glyceraldehyde-3-phosphate have been oxidized to two of phosphoglyceric acid, *making* two molecules of ATP in the process. The net energy gain so far is thus zero, but we must not forget that oxidation of glyceraldehyde-3-phosphate (reaction 6) *also* gave us two molecules of $NADH_2$. And we know from the last chapter that these can also be oxidized by the hydrogen carrier line, to give ATP in their turn.

Before the next phase of glycolysis can occur, the phosphoglyceric acid must be internally rearranged, by the transfer of the remaining phosphate from position 3 to position 2 of the molecule. The rearrangement is brought about by the enzyme *phosphoglyceromutase*:

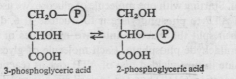

$$\begin{array}{c} CH_2O—\boxed{P} \\ | \\ CHOH \\ | \\ COOH \end{array} \rightleftarrows \begin{array}{c} CH_2OH \\ | \\ CHO—\boxed{P} \\ | \\ COOH \end{array} \qquad (9)$$

3-phosphoglyceric acid   2-phosphoglyceric acid

The tenth reaction of the series removes the elements of water from the phosphoglyceric acid, to yield phosphoenolpyruvic acid:

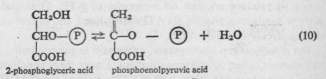

$$\begin{array}{c} CH_2OH \\ | \\ CHO—\boxed{P} \\ | \\ COOH \end{array} \rightleftarrows \begin{array}{c} CH_2 \\ \| \\ C—O \\ | \\ COOH \end{array} — \boxed{P} + H_2O \qquad (10)$$

2-phosphoglyceric acid   phosphoenolpyruvic acid

The responsible enzyme is *enolase*. (Enolase requires Mg^{++} ions as activators. Sodium fluoride, which traps magnesium by forming a complex compound with it, inhibits enolase; this is probably the most important single poisonous effect of fluoride.)

The removal of water from the phosphoglyceric acid molecule during the enolase reaction activates the phosphate bond at carbon atom 2. Thus another synthesis of ATP becomes possible, catalysed by *pyruvic phosphokinase*:

$$
\begin{array}{c}
CH_2 \\
\| \\
C-O-\text{\textcircled{P}} \quad + \quad ADP \quad \rightarrow \\
| \\
COOH \\
\text{phosphoenol-} \\
\text{pyruvic acid}
\end{array}
\qquad
\begin{array}{c}
CH_2 \\
\| \\
C-OH \quad + \quad ATP \qquad (11) \\
| \\
COOH \\
\text{(enol) pyruvic} \\
\text{acid}
\end{array}
$$

The product of this reaction, apart from ATP, is *(enol)pyruvic acid*, which changes spontaneously into the isomeric *pyruvic acid*:

$$
\begin{array}{c}
CH_2 \\
\| \\
COH \\
| \\
COOH \\
\text{(enol)pyruvic acid}
\end{array}
\quad \rightleftharpoons \quad
\begin{array}{c}
CH_3 \\
| \\
CO \\
| \\
COOH \\
\text{pyruvic acid}
\end{array}
\qquad (12)
$$

At this point we may stop for a moment and take stock of what has happened. Starting with one molecule of glucose, we used two molecules of ATP to phosphorylate it to fructose 1, 6, diphosphate. We then split the fructose into two molecules of the 3-carbon glyceraldehyde phosphate. Each molecule of glyceraldehyde phosphate in turn yielded two of ATP, one at reaction (7), the other at reaction (11). Thus for each glucose two ATPs are spent and four synthesized, a net gain of two. But we have also converted two molecules of the coenzyme NAD into $NADH_2$ (reactions 5 and 6), which can be oxidized back to NAD, and in doing so produce another *six* molecules of ATP. Thus under aerobic conditions, another six ATPs are formed, and the net gain during glycolysis is eight ATPs.

But glycolysis can also operate, as we said at the beginning of this chapter, when *no* oxygen is present. Under these conditions there has to be some other method of reoxidizing the $NADH_2$. The way the animal cell does this is by using the $NADH_2$ to *reduce* pyruvic to lactic acid, thus:

$$\begin{array}{ccc}
CH_3 & & CH_3 \\
| & & | \\
CO + NADH_2 & \rightleftarrows & CHOH + NAD \qquad (13) \\
| & & | \\
COOH & & COOH \\
\text{pyruvic acid} & & \text{lactic acid}
\end{array}$$

the responsible enzyme being our old acquaintance lactic dehydrogenase.

Reaction (13) accounts for the production of lactic acid which is observed during anaerobic glycolysis; the lactic acid is the end-product of the glycolysis reactions, and, unless oxygen is present, it cannot be metabolized further. The cell cannot afford to let large concentrations of lactic acid build up within it, though, partly because several of the glycolytic reactions are reversible, and if too much of the end product began to accumulate they would begin to run the wrong way, and partly because some means would then have to be found of coping with the changes in pH that the presence of an acid in large concentrations would provoke. The lactic acid is therefore allowed to diffuse out of the cell and to be washed away by the bloodstream.

This completes the pathway known as glycolysis; the whole sequence is summarized in Figure 16.

GLUCOSE OXIDATION

The Embden-Meyerhof pathway ends with the production of pyruvic acid; when, however, oxygen is not in short supply, the pyruvic acid is itself further oxidized to carbon dioxide:

$$CH_3COCOOH + 5(O) \rightarrow 3CO_2 + 2H_2O \qquad (14)$$

The full mechanism of this oxidation involves a whole series of reactions, and during it a considerable amount of energy is released, providing for as much as three-quarters of the cell's normal needs. Considerable experimental difficulties lay in the way of its complete unravelling, partly because a number of cofactors are involved, and partly because the process involves a *cycle* of reactants which are continually being broken down and reformed.

The cyclic mechanism is important because, as well as providing

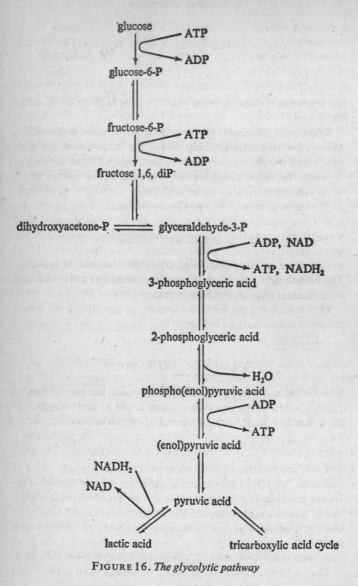

FIGURE 16. *The glycolytic pathway*

the cell with energy, glycolysis and glucose oxidation also perform another role as providers of the building blocks and intermediates for other pathways of metabolism. A cycle of metabolism enables these intermediates to be channelled to other reactions, occurring in other parts of the cell, without blocking the overall pathway by depletion. A cyclic pathway also makes it very easy for intermediates formed by the breakdown of substances other than glucose to enter and participate in the cycle, so that the cell can maximize the amount of energy it can extract from these breakdown products. However, it was this cyclic nature of the process which is so important and which for so long baffled experimenters, until the clue to its operation was provided in 1937 by Hans Krebs at Sheffield and Albert Szent-Györgyi in Budapest.

DECARBOXYLATION OF PYRUVIC ACID

The first step in the reaction sequence is the removal of carbon dioxide from pyruvic acid, and its simultaneous oxidation, in a reaction described as *oxidative decarboxylation*. This could be written

$$CH_3CO.COOH + (O) \rightarrow CH_3COOH + CO_2 \quad (15)$$

pyruvic acid $\qquad\qquad$ acetic acid

In fact, though, the product that is formed is not acetic acid at all, but a compound between acetic acid and a sulphur-containing coenzyme called *coenzyme A*. Coenzyme A, discovered by Fritz Lipmann in 1950, contains, like so many of the other coenzymes, adenine nucleotide, but, from the point of view of its reaction with acetic acid, the important part of the molecule is not the nucleotide end but the other end which terminates in an —SH group. We can thus write the coenzyme simply as CoA.SH.

The importance of the sulphydryl group is that it can easily be reduced by exchanging its hydrogen for other functional groups which it can then transfer by an energy releasing cleavage to another molecule – thus the CoA behaves, like ATP, as a group transfer molecule.

The other important feature of the pyruvic acid decarboxyla-

151

tion is that, like the other oxidations we have discussed, it operates by way of the removal of hydrogen as $NADH_2$, rather than by the participation of oxygen:

$$CH_3CO.COOH + CoA.SH + NAD \rightarrow$$
$$CH_3CO - SCoA + CO_2 + NADH_2 \quad (16)$$

pyruvic acid   acetyl-CoA.

In writing this reaction, we have not revealed the full complexity of the events surrounding pyruvate decarboxylation. The reaction sequence actually involves three separate enzymes complexed together, and three cofactors. One of these cofactors is vitamin B_1, whose chemical name is thiamine pyrophosphate (TPP). The participation of thiamine in this reaction has been suspected ever since the 1920s, when Rudolph Peters, at Cambridge, found that in animals whose diet was lacking in vitamin B_1, pyruvic acid accumulated in the blood stream, but it is only in the last few years that the exact mechanism of the reaction has been worked out. The second cofactor is another sulphur-containing compound, lipoic acid, which is alternately oxidized and reduced during the reaction. The third cofactor is, of course, CoA.

In the currently accepted reaction scheme for equation (16) pyruvic acid first combines with the thiamine pyrophosphate to yield 'acetyl-TPP' and carbon dioxide, a reaction catalysed by the first enzyme of the complex, pyruvic decarboxylase. The acetyl group is then transferred, first to lipoic acid by the enzyme lipoic acid reductase, releasing TPP, and then from lipoic acid to CoA.SH leaving the lipoic acid attached to its enzyme but in its reduced state, as dihydrolipoic acid. The third enzyme now oxidizes the dihydrolipoic acid to its original condition by effecting the transfer of its H_2 first to FAD, yielding $FADH_2$ and from thence to NAD, yielding $NADH_2$, thus completing the reaction sequence shown formally in equation (16) above. This complex sequence is probably required partly because the decarboxylation is strongly exergonic and hence must be coupled to a series of endergonic reactions in the synthesis of group transfer molecules if energy is not to be wasted. In addition, this is the first reaction we have come across in which the splitting of a carbon chain to yield the ultimate oxidative product, CO_2, occurs.

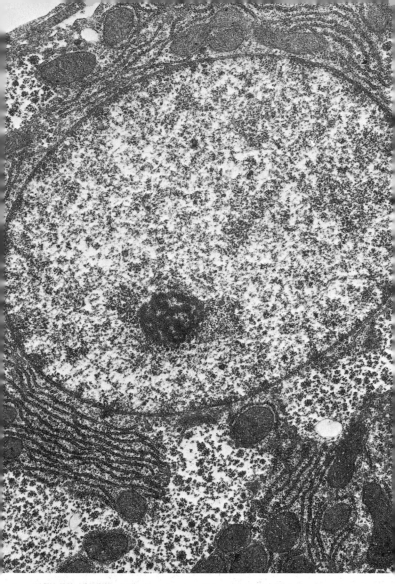

1. CELL. Rat liver cell showing nucleus with nucleolus (dark area) surrounded by nuclear membrane. Cytoplasm is criss-crossed with endoplasmic reticulum (E.R.). Attached to the E.R. are the ribosomes. The large structures are the mitochondria. Part of the cell membrane can also be seen. (× 175,000)

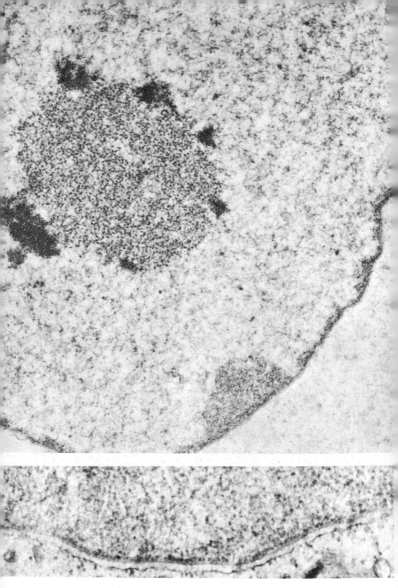

2. NUCLEUS. (*above*) Fine structure of nucleus; central portion is nucleolus. (×25,000)
    (*below*) Edge of nucleus showing double membrane structure. (×50,000)

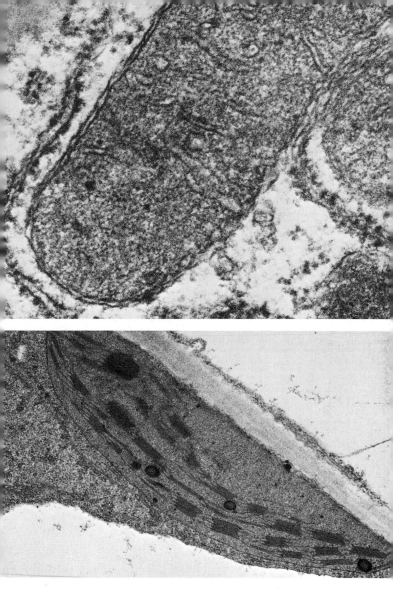

3. MITOCHONDRIA. (*above*) Showing double membrane and the internal folds or cristae. (×100,000)
CHLOROPLAST. (*below*) Showing the stacked membranous structure (grana) and starch grains. (×100,000)

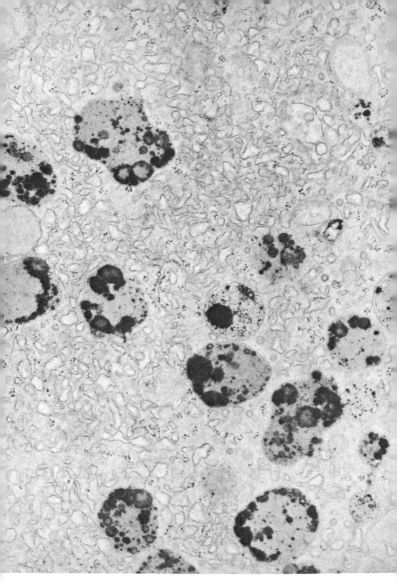

4. LYSOSOMES. Large dark objects are lysosomes within a liver cell. (×32,000)

5. MUSCLE. (*above*) Striated (voluntary) muscle, showing pattern of A, I, Z and H bands which compose contractile material. (×22,000)
(*below*) At higher power, one complete band showing thick and thin filaments. (×118,000)

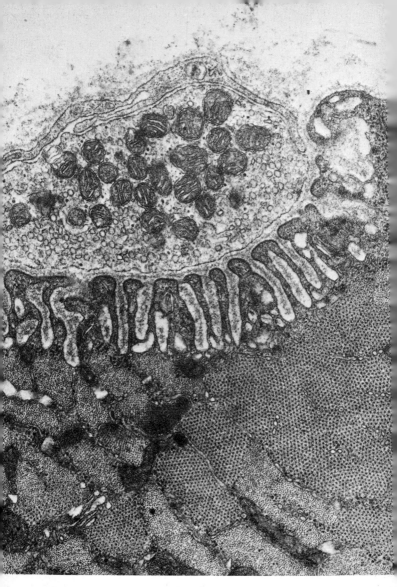

6. NEUROMUSCULAR JUNCTION. Showing the connection between nerve ending and muscle. The muscle is seen in cross section showing the bundles of myofibrils. (×20,000)

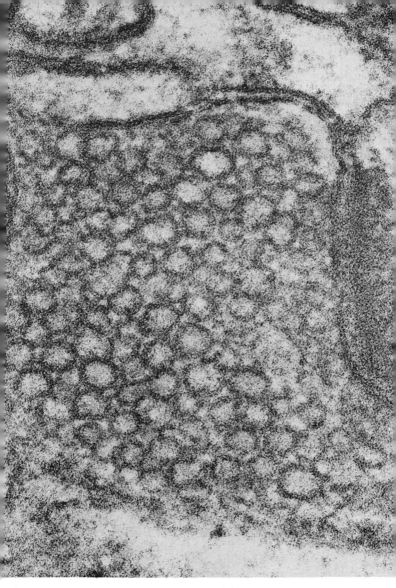

7. SYNAPSE. Junction between one nerve cell and a second is made at the darker, thicker region of membranes separating them. The presynaptic cell contains mitochondria and many, much smaller, circular vesicles, containing transmitter substance; postsynaptic cell has no such vesicles. (×300,000)

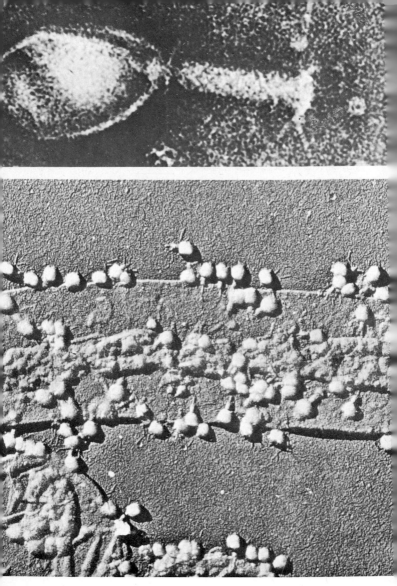

8. VIRUS. (*above*) Single virus, the bacteriophage T$_2$; a nucleic acid head surrounded by a protein jacket and 'tail'. ($\times 370,000$)
(*below*) Groups of T$_2$ bacteriophage on wall of host bacterial cell. ($\times 36,000$)

Sources of Energy

Citric acid cycle

By converting pyruvic acid to acetyl-CoA we have removed one CO_2 molecule and produced one molecule of reduced $NADH_2$. It still remains to oxidize the remaining two carbon atoms of the acetyl residue to carbon dioxide. It is here that we enter into the cycle of reactions that is associated with the name of Hans Krebs. The outline of the scheme is shown in Figure 17.

The essential feature of the cycle is that the 2-*carbon* fragment of acetyl-CoA is made to combine with a 4-*carbon* acid, to yield the very reactive 6-carbon citric acid molecule. This reaction, necessitating as it does the formation of a carbon-carbon bond, is

FIGURE 17. *The citric acid cycle*

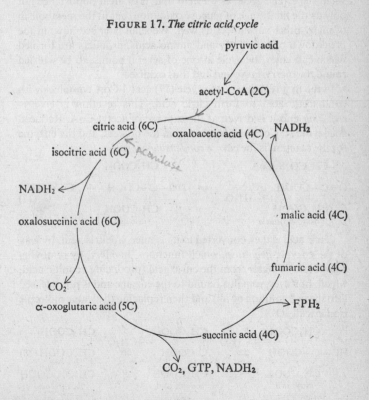

energy requiring – the energy, of course, comes from the splitting of the C_2 fragment from its group transfer molecule CoA. Citric acid is then broken down again to give first the 5-*carbon* α-oxoglutaric acid and then the 4-*carbon* succinic acid, giving off two molecules of carbon dioxide in the process. The succinic acid is then converted once more into oxaloacetic acid, and the entire cycle can start over again.

The useful thing about this merry-go-round as far as the cell is concerned is that it provides that critical feature of all desirable energy-releasing systems – a way for the acetic acid molecule to be carefully picked to pieces so as to release its potential energy in small steps. The cycle, as we shall find, is of great importance not only as the final stage of glucose oxidation, but in the metabolism of many other substances as well. We shall later see how, in the breakdown of both fatty and amino acids, products are formed which can enter the cycle at one of several points, to be whirled round the merry-go-round and thus oxidized.

In the first reaction of the cycle (17) acetyl-CoA combines with oxaloacetic acid to form citric acid. This reaction is *energy-requiring* and it is driven at the expense of acetyl-CoA. Reduced coenzyme A, CoA-SH, is released in the process, and the enzyme for the reaction is the *citrate condensing enzyme*.

$$
\begin{array}{l}
CH_3CO.S.CoA \\
\quad + \\
O{=}C{-}COOH \\
\quad | \qquad\qquad\ + \ H_2O \\
CH_2COOH \\
\text{oxaloacetic acid}
\end{array}
\rightleftharpoons
\begin{array}{l}
CH_2COOH \\
\quad | \\
HO{-}C{-}COOH \ + \ CoA.SH \\
\quad | \qquad\qquad\qquad\qquad (17) \\
CH_2COOH \\
\text{citric acid}
\end{array}
$$

Citric acid is then converted to its isomer, isocitric acid, by way of the enzyme *aconitase*, which functions, in effect, by removing a molecule of water from the citric acid (producing aconitic acid, which, however, remains bound to the enzyme and is not released into the cell solution at all) and then replacing the water molecule isomerically (18):

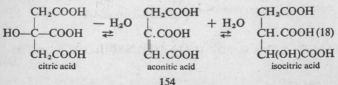

$$
\begin{array}{l}
CH_2COOH \\
\quad | \\
HO{-}C{-}COOH \\
\quad | \\
CH_2COOH \\
\text{citric acid}
\end{array}
\underset{\rightleftharpoons}{-\,H_2O}
\begin{array}{l}
CH_2COOH \\
\quad | \\
C.COOH \\
\quad \| \\
CH.COOH \\
\text{aconitic acid}
\end{array}
\underset{\rightleftharpoons}{+\,H_2O}
\begin{array}{l}
CH_2COOH \\
\quad | \\
CH.COOH \ (18) \\
\quad | \\
CH(OH)COOH \\
\text{isocitric acid}
\end{array}
$$

Isocitric acid is now *oxidized* to oxalosuccinic acid and then *decarboxylated* to give α-oxoglutaric acid. Again, both reactions are catalysed by the same single enzyme, *isocitric dehydrogenase*. In the process, a molecule of NAD is reduced to $NADH_2$. (There is also another form of isocitric dehydrogenase which uses the alternative coenzyme to NAD, NADP, and forms $NADPH_2$; the significance of this alternative enzyme will become clear in Chapter 8.) These reactions are shown in equation 19:

$$
\begin{array}{lll}
CH_2COOH & CH_2COOH & CH_2COOH \\
| & | & | \\
CH.COOH + (NAD) \rightarrow CH.COOH + (NADH_2) \rightarrow CH_2 + CO_2 \quad (19) \\
| & | & | \\
CHOH.COOH & CO.COOH & CO.COOH \\
\text{isocitric acid} & \text{oxalosuccinic acid} & \text{α-oxoglutaric acid}
\end{array}
$$

α-oxoglutaric acid resembles pyruvic acid in that it has the keto-group (CO) next to the acidic (COOH). It is thus capable of taking part in a similar reaction to the one described for pyruvic acid, that of *oxidative decarboxylation* (see equation 16). The mechanism involved is exactly similar, and demands the presence of thiamine pyrophosphate, lipoic acid, coenzyme A, and NAD. The resulting product, analogous to acetyl-CoA formed from pyruvic acid, is the group transfer molecule succinyl-CoA (20).

$$
\begin{array}{l}
CH_2COOH \\
| \\
CH_2 \quad + CoA.SH + NAD \rightarrow \begin{array}{l} CH_2.COOH \\ | \\ CH_2.CO.S.CoA + NADH_2 + CO_2 \end{array} \\
| \\
CO.COOH \\
\text{α-oxoglutaric acid} \qquad\qquad\qquad \text{succinyl-CoA} \qquad (20)
\end{array}
$$

This subsequent splitting of succinyl-CoA releasing CoA and succinic acid is used to drive a substrate level phosphorylation reaction, but this time it results in the formation, not of ATP but of another nucleoside triphosphate, guanosine triphosphate (GTP) from GDP. GTP can of course be utilized by the cell in exactly the same way as ATP, and can be converted to ATP directly at the expense of ADP, as shown below.

$$
\begin{array}{ll}
CH_2COOH & CH_2COOH \\
| & | \\
CH_2CO.S.CoA + GDP + \text{(P)} \rightleftharpoons CH_2COOH + GTP + CoA.SH \\
\text{succinyl-CoA} & \qquad\qquad (21)
\end{array}
$$

The Chemistry of Life

$$GTP + ADP \rightleftharpoons GDP + ATP \qquad (22)$$

Finally, succinic acid is converted to oxaloacetic acid by way of fumaric and malic acids, the reactions being catalysed by *succinic dehydrogenase*, *fumarase* and *malic dehydrogenase*. One molecule of reduced flavoprotein is formed at the succinic dehydrogenase catalysed step, one of $NADH_2$ by malic dehydrogenase (23):

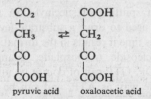

and, with the re-formation of oxaloacetic acid, the cycle is completed (see Figure 17).

It may be observed that, for the cycle to revolve, a supply of oxaloacetic acid is required with which the acetyl-CoA can react. Of course, oxaloacetic acid is being constantly re-formed as each revolution of the cycle is completed, but, nonetheless, it is not surprising to find that many cells, especially in bacteria, take precautions against running out of so essential a 'primer' by having an alternative source of oxaloacetic acid. The alternative source is provided by an enzyme, *oxaloacetic decarboxylase* which catalyses the direct 'fixing' of carbon dioxide on to pyruvic acid to form oxaloacetic acid as in equation (24):

$$
\begin{array}{cc}
\begin{array}{l} CO_2 \\ + \\ CH_3 \\ | \\ CO \\ | \\ COOH \\ \text{pyruvic acid} \end{array} \rightleftharpoons
\begin{array}{l} COOH \\ | \\ CH_2 \\ | \\ CO \\ | \\ COOH \\ \text{oxaloacetic acid} \end{array}
\end{array} \qquad (24)
$$

By a masterly piece of economy, then, many cells have a mechanism whereby the primer with which the acetyl-CoA derived from pyruvic acid must combine in order for oxidation to occur is *itself* a product of a reaction deriving from pyruvic acid.

ATP PRODUCTION

We may now try to arrive at the figure for total ATP production during the cycle. We have:

(a) pyruvic → acetyl-CoA + 1 mol. $NADH_2$	=	3 ATP
(b) isocitric → α-oxoglutaric + 1 mol. $NADH_2$	=	3 ATP
(c) α-oxoglutaric → succinyl-CoA + 1 mol. $NADH_2$	=	3 ATP
(d) succinyl-CoA → succinate	=	1 GTP*
(e) succinic → fumaric + 1 mol. FPH_2	=	2 ATP
(f) malic → oxaloacetic + 1 mol. $NADH_2$	=	3 ATP
Total		15 ATP

(* 1 GTP = 1 ATP)

Thus one molecule of pyruvic acid provides fifteen molecules of ATP.

As each molecule of glucose gives rise to *two* molecules of pyruvic acid by glycolysis, oxidation of glucose by way of the citric acid cycle yields thirty molecules of ATP. To this must be added the eight already produced between glucose and pyruvic acid and we conclude that the complete oxidation of glucose to carbon dioxide and water yields no less than thirty-eight molecules of ATP in all. As the energy released during the hydrolysis of the terminal phosphate of each molecule of ATP is about 31 kJ, the total energy trapped during the oxidation of glucose is 38 × 31 or 1,178 kJ in all. We may compare this with the equation for the complete oxidation of glucose, which (page 99) releases 2,820 kJ.

When working at maximum efficiency, therefore, the cell manages to trap and retain in usable form no less than forty-two per cent of the energy released during the burning of its fuel. This efficiency is equal to that of the most up-to-date of modern oil- or coal-fired power stations. But we must bear in mind that, like the power station, this calculation applies only to the conversion of energy in the boiler itself. The moment we come to calculate the *overall* efficiency, we have to take into account the losses of energy that occur in the transport of fuel *to* the boiler, and in the transport of energy *away* from the boiler subsequently. The overall efficiency of power-houses, or of the cell, is much lower than the figure of forty-two per cent might suggest.

OTHER SYSTEMS OF GLUCOSE OXIDATION

The Embden-Meyerhof pathway and the citric acid cycle, whilst amongst the most universal of glucose oxidation systems found in nature, are by no means the only ones. Many variants on the citric acid cycle are known to exist in different organs and species, and several alternative cycles or 'shunts' have been suggested. That the cycle itself occurs in almost all tissues has been confirmed abundantly, in particular by the use of isotopic tracers, but it is often difficult to obtain a quantitative estimate of the extent to which the glucose oxidized in the living cell uses the cycle or is diverted to other routes.

One of the most common of the alternative pathways, which may contribute anything up to thirty to forty per cent of the glucose oxidized in some tissues, is the so-called Warburg-Dickens pathway, again, in its basic outlines, the work (during the 1930s) of the biochemists after whom it is named. (Another name for the same route which is sometimes used is 'the pentose phosphate shunt'.) Like the Embden-Meyerhof route, it starts with glucose-6-phosphate, but oxidizes it and decarboxylates it in an NADP-linked enzyme system to the 5-carbon sugar phosphate *ribulose-5-phosphate*, an isomer of ribose. Six of these ribulose-5-phosphate molecules, formed from six molecules of glucose-6-phosphate, are then combined and shuffled through an intricate maze of reactions which result in the resynthesis of five molecules of the glucose phosphate once more.

The net result is that the equivalent of one molecule of glucose is oxidized to carbon dioxide and in the process thirty-six molecules of ATP are formed – two less than are given in the Embden-Meyerhof pathway coupled to the citric acid cycle. The advantage of the shunt system is that fewer enzymes are required, as there are fewer reactions involved, and also that the reactions themselves, and their enzymes, are very similar to those utilized by plants during photosynthesis.

In addition, the pathway provides pentose sugars which are

precursors for nucleic acid synthesis and a source of the reduced coenzyme $NADPH_2$. This substance is used, as we shall see, in the synthesis of fatty acids, steroids, and amino acids, and cannot be replaced by $NADH_2$ in these reactions.

It has been suggested that the shunt system evolved prior to the Embden-Meyerhof pathway, possibly before plants and animals had properly distinguished themselves from their simpler ancestors (see Chapter 13). If this were so, the Embden-Meyerhof route would represent a later, more sophisticated, development.

GLYCOGEN BREAKDOWN

A final point should be made about the entry of glucose into the oxidative systems. We have up till now regarded the glucose as entering the cell as 'free' sugar, and as being phosphorylated by hexokinase prior to oxidation. In those cells which contain their own supplies of glycogen – mainly liver and muscle – glucose is also provided by the breakdown of glycogen. An enzyme called *phosphorylase* ruptures the linkages that bind the glucose units together in the glycogen chain. The reaction requires inorganic phosphate, and glucose units are released as glucose-1-phosphate (25):

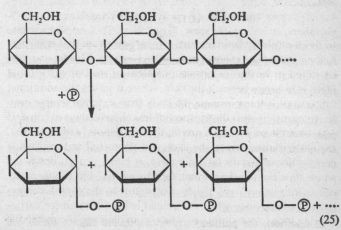

$$(25)$$

The glucose-1-phosphate is then converted by the isomerizing enzyme *phosphoglucomutase* into glucose-6-phosphate, which is in the direct path of glucose metabolism we have already discussed. We can thus draw the first stages of glucose metabolism as in Figure 18.

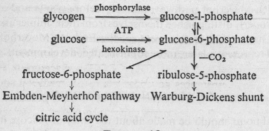

FIGURE 18.

At the entry to glucose metabolism, then, stand two enzymes, hexokinase and phosphorylase. We shall later (Chapter 11) see how, because of this strategic placing at the starting-gates to glucose catabolism, these enzymes are able to perform a critical role in *controlling* the overall rate of glucose oxidation and of regulating it to the demands of the cell for energy.

FATTY ACID OXIDATION

In times of plenty, animals turn any carbohydrate that is surplus to their immediate energy requirements into fat. This fat can then be stored in various inconspicuous areas of the body (the usual place is in layers beneath the skin, where it serves the additional function of helping insulate the body from external temperature fluctuations) against the prospect of less cheerful days to come. If and when these lean days arrive, the fat 'depots' are mobilized, the fat is transported to the liver, and there oxidized to provide energy. Some dietary fat, of course, is being broken down the whole time even in the normal, healthy animal, but it only really comes into its own as a major contributor to the overall energy-balance of the body when the animal is hard-pressed for carbohydrate, due to starvation, or when, as in diabetes, the metabolic pattern of the body becomes disturbed, so that it is unable to use glucose effectively.

Under such circumstances, the fat content of the liver begins to rise as fat from all over the body becomes concentrated there in anticipation of need, and a condition graphically described as 'fatty liver' arises; certain products of fat metabolism also begin to accumulate in the liver, the bloodstream, and the urine; these include the 4-carbon keto acid acetoacetic acid ($CH_3CO.CH_2.COOH$), and certain substances derived from it, notably β-hydroxybutyric acid ($CH_3CHOH.CH_2.COOH$) and acetone ($CH_3CO.CH_3$). These substances go under the rather medieval sounding description of 'ketone bodies', and it is their presence which represents one of the greatest dangers in diabetes, for they are poisonous.

Fat metabolism, then, has features which are of considerable medical interest, and it is perhaps for this reason that its study stretches back as far as that of glucose oxidation. The knowledge of the mechanics of the two processes has grown in parallel, the result of the work of a distinguished line of investigators, from F. Knoop in the 1900s down to the contemporary researches of Feodor Lynen in Germany, David Green and Albert Lehninger in America. We need not try to trace the many turns this research has taken, but instead can content ourselves with setting out current thinking on it.

Fats proper consist of long-chain fatty acids in ester linkage with glycerol (page 74), and the first step in their oxidation is their hydrolysis by *lipase* to release glycerol and the free fatty acids. The glycerol can be phosphorylated by ATP to give glycerol phosphate, and then oxidized by an NAD-linked enzyme to dihydroxyacetone phosphate, which, it will be recalled (page 145), lies on the direct pathway of glucose degradation in the Embden-Meyerhof scheme. From here on, glycerol follows the routes we have already mapped out to carbon dioxide and water (26):

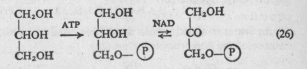

$$\begin{array}{ccccc} CH_2OH & & CH_2OH & & CH_2OH \\ | & \xrightarrow{ATP} & | & \overset{NAD}{\rightleftharpoons} & | \\ CHOH & & CHOH & & CO \\ | & & | & & | \\ CH_2OH & & CH_2O{-}\textcircled{P} & & CH_2O{-}\textcircled{P} \end{array} \qquad (26)$$

The Chemistry of Life

The problem of fat oxidation then resolves itself into that of the fate of the fatty acids derived from the fat. These, the straight chain acids stearic (18-carbon), palmitic (16-carbon), and oleic (18-carbon, one double bond), are oxidized, according to experimental results obtained by Knoop at the turn of the century, by a process which splits off the carbon atoms two at a time; thus stearic is converted first to palmitic, then to a 14-carbon acid, then a 12-carbon acid, then a 10-carbon, and so on, until the molecule is whittled right down to the 4-carbon acetoacetic acid, which is split in its turn into two 2-carbon acetic acid fragments. The removal of carbon atoms two at a time like this is called β-oxidation, as it occurs by the oxidation of the carbon atoms two places away from (in chemical parlance, in the β-position to) the acidic group at the end of the fatty acid chain.

The first reaction in the fatty acid oxidation sequence is the thioesterification of the acid by CoA-SH. This involves bond formation between the CoA-S and the fatty acid and is therefore endergonic; it is achieved by linking it to the simultaneous hydrolysis of ATP to AMP. The enzyme which does the job is a *thiokinase* (27).

$$R.CH_2.CH_2.CH_2COOH + CoA.SH + ATP$$
$$\rightleftharpoons R.CH_2.CH_2.CH_2CO.S.CoA + AMP + 2\textcircled{P} \quad (27)$$

Thioesterification takes place within the cytoplasm but the rest of the reactions which constitute β-oxidation occur within the mitochondria. However, fatty acyl CoA cannot penetrate the mitochondrial membrane, so for the transport to occur, the CoA group is substituted by a compound called carnitine, which carries the fatty acyl group into the mitochondria. Once inside it reverts to acyl CoA.

Acyl-CoA is now oxidized by a flavoprotein dehydrogenase enzyme (28):

$$R.CH_2CH_2CH_2CO.SCoA + FP$$
$$\rightleftharpoons R.CH_2CH=CH.CO.SCoA + FPH_2 \quad (28)$$

and the water is added across the double bond so produced by a *lyase*, yielding β-hydroxyacyl-CoA (29):

162

$$R.CH_2CH=CH.CO.SCoA \quad + \quad H_2O$$
$$\rightleftarrows \; R.CH_2CH.CH_2CO.SCoA \quad (29)$$
$$\qquad\qquad\qquad\quad | $$
$$\qquad\qquad\qquad\quad OH$$

Another dehydrogenase, this time NAD-linked, converts the hydroxy- to the β-ketoacyl-CoA (30):

$$R.CH_2CH.CH_2CO.SCoA \quad + \quad NAD$$
$$\;\;| $$
$$\;\;OH$$
$$\rightleftarrows \; R.CH_2C.CH_2CO.SCoA \quad + \quad NADH_2 \quad (30)$$
$$\qquad\qquad\quad || $$
$$\qquad\qquad\quad O$$

This keto acid is enabled to react with another molecule of coenzyme A (31) by *β-ketothiolase*:

$$R.CH_2C.CH_2.CO.SCo.A \quad + \quad CoASH \; \rightleftarrows \; R.CH_2CO.SCoA$$
$$\;\;\; || $$
$$\;\;\; O$$
$$+ \quad CH_3.CO.SCoA \quad (31)$$

This last reaction succeeds in *splitting* the carbon chain of the acid, releasing acetyl-CoA and the CoA derivative of the fatty acid with *two fewer* carbon atoms than the original molecule of equation (27). This reaction is known as *thiolysis*. In it, a molecule of CoA is added to the fatty acyl residue without the need of energy from ATP. The thiolysis reaction itself provides enough energy for the addition to be made without need of an external energy source as well. As well as releasing a molecule of acetyl-CoA, the thiolysis reaction therefore also produces a fatty acyl-CoA which can now take part in the same reaction sequence that we have already traced, beginning with reaction (28) above. That is, each time that a fatty acyl-CoA molecule passes through the sequence of reactions (28) to (31), of oxidation, hydration, oxidation, and thiolysis, it produces a daughter molecule of acetyl-CoA and an acyl-CoA molecule two carbons shorter than its parent but ready in its turn to breed in the same manner.

The fatty acid molecule thus undergoes a repetitive series of four reactions which steadily chip away at its carbon skeleton, releasing the fragments as acetyl-CoA. We can show the complete process for the 6-carbon *hexanoic acid* like this:

$$CH_3CH_2CH_2CH_2CH_2COSCoA$$

$$CH_3CH_2CH_2CO.SCoA \quad + \quad CH_3CO.SCoA$$
$$+ \quad NADH_2 \quad + \quad FPH_2 \quad (32)$$

$$CH_3CO.SCoA \quad + \quad CH_3CO.SCoA$$
$$+ \quad NADH_2 \quad + \quad FPH_2$$

and it then appears that from hexanoic acid we obtain three molecules of acetyl-CoA, two of $NADH_2$, and two of FPH_2. When the acetyl-CoA is oxidized through the citric acid cycle, each molecule gives twelve of ATP; the two molecules of $NADH_2$ are worth six ATPs, and the two flavoproteins another four. The yield of ATP produced by the complete biological oxidation of hexanoic acid is thus thirty-six plus 6 plus four or forty-six ATPs in all; as two are expended in forming hexanoyl-CoA from the hexanoic acid we began with, the net gain in ATP is forty-four molecules. This may be compared with the thirty-eight produced during the oxidation of glucose. The 6-carbon fatty acid thus releases *more* usable free energy than the 6-carbon carbohydrate.

However, the oxidation is probably not as efficient as it looks at first sight, for we have assumed in the calculation that all the acid is broken down into acetyl-CoA which is subsequently oxidized. But in practice, we have already noted that under conditions where fatty acids form the major energy source of the body, during starvation or diabetes, large amounts of the substance acetoacetic acid accumulate. Now one source of acetoacetic acid is the combination of two of the molecules of acetyl-CoA formed during fatty acid oxidation (33):

$$CH_3CO.SCoA \quad + \quad CH_3CO.SCoA$$
$$\rightleftarrows \quad CH_3COCH_2CO.SCoA \quad + \quad CoASH \quad (33)$$

Evidence from isotopic tracer experiments suggests that acetoacetic acid is in fact formed in this way during fatty acid oxidation.

However, every molecule of acetoacetic acid or its derivatives that the liver accumulates and subsequently excretes means that two potential acetyl-CoA molecules have not been exploited so as to release their potential energy through oxidation by way of the citric acid cycle. Why? Particularly as fatty acid oxidation only occurs under conditions of carbohydrate starvation, one may well be surprised at the fact that the cell seems to squander a valuable source of energy in this wasteful way. To discard acetoacetic acid like this is to throw away fuel before it is more than half-burned.

The reason seems to be tied up with the way the citric acid cycle works. We mentioned earlier that, in order for the cycle to turn, acetyl-CoA must combine with oxaloacetic acid to form citric acid, and that, although in theory only catalytic amounts of oxaloacetic acid are required, since it is constantly re-formed as the cycle comes full circle, nonetheless the cell in practice seems to find it necessary to provide an additional source of oxaloacetic acid by way, for instance, of reaction (24) of page 156. In such reactions, oxaloacetic acid is produced from pyruvic acid. But pyruvic acid lies only on the pathway of glucose oxidation, and not on that of fatty acid oxidation. There is no pyruvic acid at all formed during the series of reactions we have just described. It follows that the supplies of oxaloacetic acid must be sharply diminished during carbohydrate starvation. Thus the citric acid cycle must gradually slow down, as a queue of acetyl-CoA molecules is formed, lining up and waiting to be linked to oxaloacetic acid and taken for a ride on the merry-go-round.

And as the amount of acetyl-CoA increases, an increasing number of molecules will tend to combine with one another by reaction (33). It seems then that, in the absence of carbohydrate, the cell, forced to turn for free energy to its fat supplies, finds that it cannot even use these as efficiently as would be possible were carbohydrate present to provide the essential oxaloacetic acid. This represents a strange weakness and inefficiency on the part of the cell, and serves to emphasize once again that, so far as energy is concerned, there is no substitute for sugar.

OTHER SOURCES OF ENERGY

Fats, we have seen, are used as an energy-source only when the animal is starved of glucose. If starvation is prolonged, though, the body's stocks of fat begin to run down and finally are exhausted. It becomes necessary to search elsewhere for oxidizable substrate. The only remaining substances present in the body which can provide this last resort are the proteins. The proteins are so essential to survival that it is only with the greatest reluctance that the cell will turn to them as an energy source. Yet, if all other supplies fail, ultimately they too begin to be consumed to stave off the day when no more life-preserving ATP can be produced. It is as if, having burned all other fuel, one began in desperation to chop up the furniture and throw it too on the fire.

Obviously, the process cannot be continued indefinitely; under normal circumstances, the body maintains a 'nitrogen balance', the amount of nitrogen being taken in from the intestines as amino acids being exactly equivalent to the amount that is released during the breakdown and disposal of 'worn-out' protein and excreted as urea in the urine. As starvation proceeds, the nitrogen balance is tipped – increasing amounts of nitrogen are excreted as urea as the proteins of which it formed part are oxidized for energy, and the body begins to break down its protein more rapidly than it replaces it. The amounts of nitrogen excreted rise day by day and finally shoot up dramatically, as if the body despaired of being able to conserve any protein at all, and this last great rise is rapidly followed by death.

Apart from an understandable reluctance to waste good protein as an energy source, its use by the cell is complicated by the difficulties attached to providing for the safe disposal of the nitrogen contained in the protein. Useless to animals as a source of energy, the nitrogen tends to accumulate as ammonia, NH_3. But ammonia is dangerously poisonous; even very small amounts injected into the bloodstream can be lethal. The body is therefore committed to finding a non-toxic disposal system for the ammonia which protein metabolism produces. This it does by converting the ammonia into the harmless urea, $CO(NH_2)_2$, which can be passed through the bloodstream to the kidneys and thence excreted into the urine without fear of self-poisoning.

The fact that the urine contains a fair amount of urea even while the body is healthy and functioning normally indicates that protein breakdown is occurring all the time; we have already seen how the proteins, along with other of the giant molecules which the cell produces, are constantly being destroyed and re-formed in the ceaseless self-renewal which the cell undertakes and which is described as the 'dynamic state of body constituents'. Obviously, the cell regains what energy it can during this breakdown process, and so, like the fats, amino acids and proteins contribute to the energy-balance of the body even when carbohydrate is plentiful. Under normal circumstances, though, their services to this cause must be very slight.

The conversion of protein nitrogen into urea is a process that takes place almost exclusively in the liver (dogs whose livers have been removed can live for some time on a nitrogen-free diet, but quickly die if fed protein as they now have no means of preventing nitrogen being released as ammonia and poisoning them). The mechanism of urea formation was largely discovered by Hans Krebs in the 1930s; like that of acetyl-CoA oxidation, it is a *cyclical* process of considerable theoretical interest. But as it is not a source of energy to the cell, we shall not describe it further here.

Such energy as the protein can provide comes through the oxidation of the carbon skeletons of its constituent amino acids. As there are many amino acids, with quite widely differing structures, it is hardly surprising that there are several different oxidative routes available. In all of them, though, the principle of the degradation is similar; the amino acid is disembarrassed of its amino-nitrogen group at an early stage and the denuded molecule is converted into substances which lie on the direct pathway of either glucose or fatty acid oxidation, where it can be disposed of by reaction pathways we have already described. Amino acids which give rise to glucose oxidation and citric acid cycle intermediates are known as *glucogenic*, whilst those which are converted into substances lying on the pathway of fatty acid metabolism are *ketogenic*.

There are several ways of ridding the amino acid of its nitrogen. One such would be direct oxidation, in which the acid is first dehydrogenated, passing two hydrogens to an acceptor, and then hydrolysed by water, releasing ammonia (34, 35):

$$\begin{array}{ccc} \text{R.CHCOOH} & \rightleftharpoons & \text{R.C.COOH} + 2\text{H} \\ | & & \| \\ \text{NH}_2 & & \text{NH} \end{array} \qquad (34)$$

amino acid   imino acid

The Chemistry of Life

$$R.C.COOH + H_2O \rightleftharpoons R.CO.COOH + NH_3$$
$$\underset{\text{imino acid}}{\overset{\parallel}{NH}} \hspace{4cm} \underset{\text{keto acid}}{}$$

$$(35)$$

The other product of reaction (35) is a keto acid, whose further fate depends on the exact nature of the R-group in it. Reactions of the type of (34) are catalysed by amino acid oxidases. The second reaction (35) is spontaneous and requires no enzyme. The most important of the amino acid oxidases is *glutamic dehydrogenase*, which converts glutamic into α-oxoglutaric acid (generating $NADH_2$ in the process) (36):

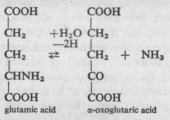

$$(36)$$

The α-oxoglutaric acid is of course a member of the citric acid cycle series of acids, and we have already described its fate.

The importance, though, of the glutamic dehydrogenase reaction is that, apart from this enzyme, the only amino acid oxidases present in the cell are either rather weak enzymes, or else are ones that are specific not for the naturally occurring (−)isomers of the amino acids, but for the unnatural (+)isomers. In general, amino acids other than glutamic have a slightly different method of freeing themselves of nitrogen. This alternative route is called *transamination*, and transaminase enzymes exist for many amino acids so that, starting with any other amino acid and α-oxoglutaric acid, the exchange reaction (37) can occur:

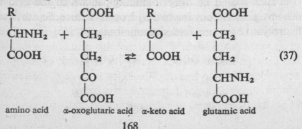

$$(37)$$

Under the influence of transaminase, the amino acid swaps its NH_2 amino group for the keto group of α-oxoglutaric acid, releasing the appropriate keto acid and glutamic acid. The keto acid can go on its own metabolic way by way of reactions we have already seen in the glycolytic and glucose oxidation pathways, and glutamic acid can be deaminated by glutamic dehydrogenase to produce α-oxoglutaric acid once more. By this means reactions (36) and (37) can be coupled to provide yet another of the now familiar cyclical processes in which the cell delights (Figure 19). This way, virtually all amino acids are sucked into the metabolic whirlpool and made to yield their potential energy.

FIGURE 19.

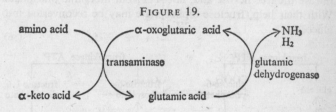

INTERCONVERSIONS OF FATS, AMINO ACIDS, AND CARBOHYDRATES

All the cells of the body metabolize glucose. But fatty acids and amino acids are utilized mainly in the liver, although some amino acid metabolism can also occur in other tissues, such as brain or kidney. When starvation conditions arise, although the liver can get all the energy it needs by burning fat, other tissues would tend to go short if there were no means of dispatching energy to them. The most convenient way of sending packets of energy over long distances in the body is to parcel it up as glucose. So the liver needs to be able to convert the foodstuffs it uses into sugar for the benefit of less versatile tissues. What in fact this demands is that the liver cell provide a system of reversing the sequence of reactions that leads from glucose to acetyl-CoA, for both fats and amino acids produce this substance during their oxidation.

Fortunately, nearly all of the reactions that we have traced out in describing the pathway of glucose oxidation are reversible. Given the right conditions, all the enzymes which help to break down fructose 1, 6, diphosphate to pyruvic acid (reactions 4 to 12) can be made to retrace their steps. Fructose 1, 6, diphosphate itself is made from glucose with the help of two molecules of ATP and the enzymes hexokinase, phosphofructokinase, and phospho-hexoisomerase (reactions 1 to 3, page 144). The two ATP molecules which are spent in 'priming' the hexose prior to breaking it down are irrevocably lost, for both the hexokinase and phos-phofructokinase reactions are irreversible. However, there exist phosphatase enzymes which can hydrolyse hexose phosphates to release the free hexose and, not ATP, but inorganic phosphate. With their help, fructose diphosphate may be reconverted into glucose:

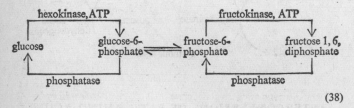

(38)

Thus pyruvic acid and glucose are connected by a reversible se-quence of reactions. But fats and amino acids yield either acetyl-CoA or citric acid cycle intermediates like α-oxoglutaric acid, not pyruvic acid. In the glucose oxidation scheme, pyruvic acid is oxidatively decarboxylated to acetyl-CoA (equations 15 and 16, pages 151, 152) by a complex reaction sequence that appears to be virtually irreversible. In order to resynthesize pyruvic acid, we need to make use of a more roundabout method. The reactions of page 156 showed one way in which, starting with oxaloacetic (or malic) acid, some types of cell can resynthesize pyruvic. As both these substances are formed during the rotation of the citric acid cycle, both acetyl-CoA and the cycle intermediates can give rise to pyruvic acid. Although the details of the exact ways in which these reactions actually occur are by no means yet fully mapped out, and it is not certain whether all cells of all species contain this mechanism, it seems clear that there *are* ways in which the cell can, at a pinch, regain glucose at the expense of other foodstuffs. Equally, glucose itself can be used as the starting substrate for the synthesis of fatty acids, amino acids, fats, and proteins that is discussed in the next chapter. We can summarize all these interrelationships in one comprehensive diagram, Figure 20.

FIGURE 20. *Interrelation of fats, carbohydrates, and proteins*

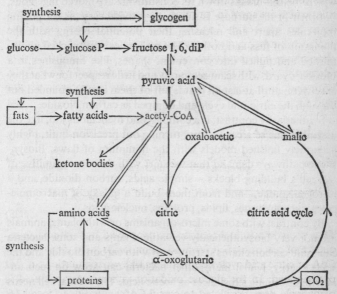

CHAPTER 9

THE SYNTHETIC PATHWAYS

It is notoriously easier to destroy than to build. The methods by which the cell breaks down its constituents are indeed ingenious, but their prime purpose is simply to provide systems for pulling molecules apart and releasing their potential energy with the minimum of fuss and bother. The molecules are knocked about, pushed and pulled into convenient shapes, like limousines in a breaker's yard, with scant ceremony and little respect for what they once were, until at last all that is left of them can be pounded out through the citric acid cycle and dumped as carbon dioxide.

Synthesis, by contrast, is a complex and meticulous job, like the assembly-line of a car factory, turning out precision-built, highly accurately finished models with the minimum of flaws. Biosynthetic pathways demand that we start with the most primitive of the cell's building blocks – simple acids, carbon dioxide, and a nitrogen-source – and from them build a galaxy of macromolecules, carbohydrates, lipids, proteins, nucleic acids.

By contrast with some micro-organisms or with plants, animals are not very biosynthetically versatile. Plants and some bacteria can build carbon chains starting only with carbon dioxide and the sun's energy, whilst many other bacteria can grow on such un-nutritive-seeming substances as acetic acid. The animal cell needs much more; before it can set to work it must be supplied with sugar, several different sorts of amino acid, and a variety of vitamins. In truth, the animal lives parasitically on the ability of others to synthesize the substances it needs; freed from the necessity of staying in one place with its arms out to catch the sun's rays, or of touring the countryside for a supply of acetic acid, it can instead do other, more interesting things. Even so, one mustn't underrate the skill with which even the backward cells of our own bodies set about making proteins or lipids; they still do better than the scientist knows how to imitate.

But even though there are differences in ability between plants and animals when it comes to synthesis, many general features are possessed in common, and a large number of the more important of pathways are shared by all living organisms. Certain properties characterize all biosynthetic processes. Firstly, just as the catabolic systems we described in the last chapter are *providers* of energy, so biosynthetic routes *use* energy.

Much of the ATP produced during oxidation of glucose or of fatty and amino acids is used in the formation of highly reactive phosphorylated intermediates on the pathways for the building of new molecules of carbohydrates, fats, and proteins.

We have already pointed out that these macromolecules are being constantly renewed by the cell, so that even the healthy, fully-grown animal is kept busy the whole time making new molecules. It is difficult if not impossible to assess just how much of the cell's ATP is being used for this purpose. But it is possible to calculate that in most active cells every molecule of ATP is sapped of its energy, converted into ADP, and resynthesized to ATP again at the expense of oxidizable substrate many times a minute. And we can also be sure that however efficient the utilization of ATP is, one hundred per cent efficiency in biosynthesis will not exist any more than one hundred per cent of the energy released during biological oxidation could be trapped and exploited by the cell; indeed, we shall see how the cell nearly always has to expend *more* molecules of ATP in making any particular substance than could be gained by oxidizing it. This, of course, is why we die without food – we need a constant supply of potential energy from outside our own bodies just to keep ourselves ticking over, even before we begin to think about actually doing work on our surroundings.

All biosynthetic pathways, then, are energy-requiring, and in this sense are the reverse of those catabolic pathways we have so far discussed. Many biosynthetic pathways are the reverse of catabolic ones in another sense also. The energy-giving reactions we have described proceed by oxidation of their substrates; biosynthesis, on the other hand, demands the *reduction* of its simpler starting molecules as the more complex materials are built up. Just as cellular oxidation was mediated by coenzymes, and in particular by NAD, we shall find that cellular reduction is also coenzyme demanding, and that the key hydrogen donor is, although not $NADH_2$, its close relative, the phosphorylated

NADPH$_2$, which we have already met in the role of a hydrogen *acceptor* for a small number of oxidative dehydrogenase reactions.

The reaction NADPH$_2$ → NADP has a $\Delta G°$ of -167 kJ so its use as a hydrogen donor also provides the means by which this exergonic reaction can be linked to an otherwise endergonic reaction to ensure an overall negative free energy change. This then is another molecule behaving in an analogous way to ATP in the cellular energy economy.

Biosynthesis, as should now be apparent, is a more complicated process than oxidation, and is by no means yet completely unravelled. The problems of investigating biosynthesis are complicated by the fact that many of the catabolic reactions discussed in the last chapter are reversible – we have already seen how, starting with acetyl-CoA or pyruvic acid, the cell can synthesize glucose by a reversal of the Embden-Meyerhof pathway. It was thus a fairly natural assumption that biosynthetic routes in general occurred by the exact reversal of oxidative ones, and many distinguished investigators, especially in the fields of glycogen and of fat synthesis, were led astray by failing to recognize that though an enzyme might in its purified form be *able* to catalyse both a forward and a backward reaction in the test-tube, in the cell it is not likely ever to be given the freedom to do this. Conditions and relative concentrations of the reactants are always likely to favour one direction rather than the other. It would indeed be a very false economy for the cell to attempt both to degrade and synthesize molecules using the same set of enzymes. The confusion that would result would be like that in a busy shop where crowds of people try to climb a staircase that is already full of shoppers coming down.

Life is easier with two stairways, marked clearly 'up' and 'down'. It has become clear within the last five or six years that the cell does the same. Let us look at some typical examples of 'up' staircases, for the synthesis of glycogen, fats, and some of the more complex small molecules the cell uses. We will leave the most complex of the up staircases, those for protein and nucleic acid synthesis, to a chapter of their own.

The Synthetic Pathways

SMALL MOLECULES

The last chapter has shown how the metabolism of small molecules like glucose and the amino acids interacts. Certain cells, though not all, can resynthesize glucose from intermediates of the tricarboxylic acid cycle, fixing carbon dioxide in order to do so. Amino acids, too, can be generated from the intermediates of glycolysis and the tricarboxylic acid cycle, although a significant proportion of the amino acids present in protein cannot be synthesized, at least in humans, who lack the necessary enzymes. Hence, such amino acids as tryptophan, phenylalanine, leucine, valine, methionine, lysine, threonine, and isoleucine must be supplied in the diet (so-called essential amino acids). Serine, glycine, and cysteine can be made from 3-phosphoglyceric acid, provided glutamic acid is present as a donor of NH_2 groups. The glutamic acid itself can be made by a reaction between α-oxoglutaric acid (a tricarboxylic acid cycle intermediate) and ammonia, and participates in the formation of aspartic acid, arginine, proline, and other amino acids. Aspartic acid and glutamic acid also provide the nitrogen for the production of purines and pyrimidines, while the sugar portion of the nucleotides comes by way of ribose 5-phosphate, itself a product of the pentose phosphate pathway. Important and interesting as these metabolic interactions and syntheses are, we will not consider them in any further detail here (they form a large part of the central chapters of many standard text-books of biochemistry) but pass on straight away to a consideration of the synthesis of the macromolecules.

GLYCOGEN SYNTHESIS

The glucose polymer of animals, glycogen, is made mainly by liver and muscle cells. The starting material for its synthesis may be regarded as any glucose surplus to the immediate energy requirements of the cell, and we can view the glycogen as being made by the binding together of glucose units which have been activated by phosphorylation from the ATP molecule. We previously traced the breakdown of glycogen to provide oxidizable glucose at the hands of the enzyme phosphorylase (page 159), which, beginning

175

with a chain of glucose units and inorganic phosphate, can split the molecules off from the chain one at a time as glucose-1-phosphate. The phosphorylase reaction is reversible, and the purified enzyme, if provided with glucose phosphate and a 'primer' of a bit of glycogen chain (it needn't be more than two glucose units long), will perform in the test-tube the reaction of *glycogen synthesis* – it will begin sticking glucose units on to the primer chain and releasing inorganic phosphate. For a long time it was thought that this really was the way glycogen was made within the cell itself, despite the fact that the equilibrium of the phosphorylase reaction in fact lies far in the direction of glycogen breakdown and not that of synthesis, so that even in the artificial, test-tube situation the glucose phosphate must be present in very high concentrations to drive the synthetic reaction.

It was not until the 1960s that it became quite clear that, in the living cell, glycogen synthesis followed quite another and more complicated route than the reversal of phosphorylase action, and that in this new pathway the equilibrium of the reaction was shifted decisively away from the sugar phosphate and towards the polymer chain formation. This breakthrough followed the discovery by Luis Leloir in Argentina of a new coenzyme derived from the pyrimidine, uracil.

This substance, uridine diphosphate, UDP, can participate in a group transfer reaction with ATP to form the triphosphate, UTP, and it is UTP that is the common intermediate that provides the coupling link between ATP breakdown and the energy-requiring build up of the glycogen chain. UTP, Leloir found, can react with glucose-1-phosphate to give *uridine diphosphate glucose,* or UDPG for short (1):

UTP + glucose-1-phosphate → UDPG + inorganic pyrophosphate (1)

Glucose, by being converted into glucose-1-phosphate, has already been activated. Now, when it combines with UTP to form UDPG it becomes very reactive indeed. Attached to its UDP coenzyme, the glucose can undergo a whole series of conversions that are otherwise impossible for it; it can, for example, be readily converted into several other hexose isomers of glucose, such as galactose. But from our point of view the most important reaction of UDPG is that catalysed by the enzyme glycogen synthetase. This enzyme easily performs the reaction which phosphorylase finds so hard, and transfers the glucose from UDP on to a 'primer' to begin building the glycogen chain:

$$\text{UDPG} + \underset{\text{(primer)}}{\text{gluc-gluc}} \rightarrow \text{UDP} + \text{gluc-gluc-gluc} \qquad (2)$$

$$\text{UDPG} + \underset{\text{etcetera}}{\text{gluc-gluc-gluc}} \rightarrow \text{UDP} + \text{gluc-gluc-gluc-gluc} \quad (3)$$

This reaction can be repeated indefinitely until the chain at last reaches the right length. Meanwhile, the UDP set free is converted into UTP at the expense of ATP:

$$\text{UDP} + \text{ATP} \rightleftarrows \text{UTP} + \text{ADP} \qquad (4)$$

Two ATP molecules are therefore consumed for each glucose unit added to the glycogen chain, and not the *one* which would be the case if the synthesis were in fact catalysed by phosphorylase.

This does not quite complete the story of glycogen synthesis, because, as we know (page 54), the glycogen molecule is not straight-chained but *branched*, and both the UDPG system and the phosphorylase one are only capable of polymerizing glucose into straight chains linked by 1–4 bonds between the glucose units. The branching points are, it will be recalled, made by joining one length of 1–4 linked chain to another by a 1–6 linkage.

The links are provided by the action of the enzyme amylo (1,4–1,6) transglycolase which works by a highly specific mechanism so as to form the branch points in exactly the same way each time. The enzyme removes a fragment of six or seven glucose residues from the non-reducing end of a linear chain (that is, the end that has a free hydroxyl at C4 rather than C1) and attaches the fragment to the hydroxyl group on C6 of any glucose molecule in the third place in from the non-reducing end of another chain. The repeated concerted action of glycogen synthetase and the transglycolase enzyme results in the formation of the very characteristic highly branched glycogen molecules each of which contains about 200 to 300 glucose units in all, to be deposited as short-term food stores in the cells of the liver and muscle.

Glycogen synthesis in many ways provides a general model for polysaccharide biosynthesis because in every one of these pathways so far studied nucleoside diphosphates like UDP seem to act as the glycosyl carrier, donating glycosyl molecules to an incomplete polysaccharide chain.

FATTY ACID SYNTHESIS

The problem of fatty acid synthesis, like that of glycogen, is an example of biochemists being led astray, by the relative ease with which the oxidative pathway for fatty acid breakdown was unravelled, into believing that the synthetic route was simply its reverse. The oxidation of fatty acids by the enzyme system of pages 160–165 proceeds by way of alternate hydration and dehydrogenation of the coenzyme A derivatives of the acids, in a sequence of repeating reactions which chip away 2-carbon fragments from the end of the carbon chain and release them as acetyl-CoA. All the reactions in the sequence are reversible, at least in the test-tube, and up till only a few years ago it was believed that this reversal was the general mechanism responsible for the manufacture of the fatty acids in the living cell.

Certain important differences have come to light, though. For instance, the oxidative pathway uses the coenzyme NAD, whilst when the purified enzymes of the synthetic system were studied, it was found that they worked better with $NADPH_2$ as coenzyme than with $NADH_2$. So it was proposed that the synthetic system was the reversal of the oxidative system with one difference – that NADP replaces NAD at the critical reduction step:

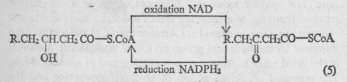

$$R.CH_2 \underset{\underset{\displaystyle OH}{|}}{CH}.CH_2 CO—S.CoA \qquad\qquad R.CH_2 \underset{\underset{\displaystyle O}{\|}}{C}.CH_2 CO—SCoA$$

oxidation NAD

reduction $NADPH_2$ (5)

But this proved to be only the first of a set of modifications which had to be made for effective synthesis of fats. It was, for instance, observed that cell preparations which had been made in solutions which contained the bicarbonate ion $HCO_3{}^-$, a ready source of dissolved carbon dioxide, made fatty acids much better and faster than those that were made in non-bicarbonate solutions. This immediately suggested that in some way fatty acid synthesis required carbon dioxide, yet, when experiments were made using radioactive CO_2, none of the label appeared in the fatty acids. The effect of the CO_2 appeared to be a catalytic one. Most important, though, was a theoretical problem similar to that of glycogen synthesis. The last of the reactions of fatty acid oxidation is that of *thiolysis* by β-ketothiolase (6):

$$R.COCH_2CO.SCoA \;+\; HSCoA \;\rightleftarrows\; RCH_2CO.SCoA \\ +\; CH_3CO.SCoA \qquad (6)$$

(see also equation (31), page 163).

Like the phosphorylase reaction in glycogen metabolism, although thiolysis is theoretically reversible, the equilibrium lies far over to the right; in order for the cell to use it to build fatty acids it has to collect a formidably high concentration of acetyl-CoA and $RCH_2CO.SCoA$. Again, a reaction possible in the test-tube is by no means so easy for the cell in practice. It is clear that all these facts, taken together, point to the existence of an alternative route of fatty acid synthesis.

The clue to the new pathway was provided in 1958 by S. J. Wakil, in the U.S.A. Wakil sought a system whereby the cell could bypass the unfavourable equilibrium of reaction (6), and found that the trick was done, just as by UDPG in glycogen synthesis, by an activation of the reactants. Once again, activation is provided at the expense of ATP and a new coenzyme. What the cell does is to convert acetyl-CoA into another, more reactive substance, *malonyl-CoA*. Malonic acid is produced by the addition of a molecule of carbon dioxide to acetic acid, (7):

$$CH_3COOH \;+\; CO_2 \;\rightarrow\; CH_2\begin{smallmatrix}COOH\\[2pt]\\[2pt]COOH\end{smallmatrix} \qquad (7)$$

and it is obvious from its formula that the molecule will be more unstable, and thus more reactive, than that of acetic acid.

The enzyme responsible for carbon dioxide fixation is called acetyl-CoA carboxylase and is dependent upon the presence of the metal ion manganese and also the vitamin biotin. In fact it is the biotin that first reacts with the CO_2 in an energy-requiring reaction that is driven by the simultaneous breakdown of ATP. The carboxybiotin which is formed can now be thought of as an activated carrier molecule, very similar to UDPG. It can easily transfer its CO_2 to acetyl CoA to give malonyl CoA in the following reactions:

$$\text{Biotin} + CO_2 + \text{ATP} \rightarrow \text{carboxybiotin} + \text{ADP} + \text{\textcircled{P}} \qquad (8)$$

$$CH_3CO.SCoA + \text{carboxybiotin} \rightarrow \underset{\overset{|}{CO-S.ACP}}{\overset{COOH}{\overset{|}{CH_2}}} + \text{biotin} \qquad (9)$$

The next step in the pathway is the linking of malonyl-CoA and acetyl-CoA to different molecules of another sulphydryl containing protein, acyl carrier protein (ACP). It is on their ACP derivatives that malonyl-CoA and acetyl-CoA combine to give the β-keto acid acetoacetic acid, still linked to ACP, but releasing CoA and CO_2.

$$CH_3-\overset{\overset{\displaystyle O}{\|}}{C}-S.ACP + \underset{\underset{\displaystyle CO-S.ACP}{|}}{\overset{\overset{\displaystyle COOH}{|}}{CH_2}} \rightarrow$$

$$CH_3-\overset{\overset{\displaystyle O}{\|}}{C}-CH_2-\overset{\overset{\displaystyle O}{\|}}{C}-S.ACP + ACP-SH + CO_2 \qquad (10)$$

Thus carbon dioxide is alternately fixed by reaction (8) and released once more by reaction (10); it is in fact acting catalytically, as we should expect from the experiments made in the presence of radioactive bicarbonate that we have already mentioned.

The net results of reactions (8)–(10) is the linking of two molecules of acetyl-CoA to give one of acetoacetyl-CoA. As the equilibrium of the reactions is favourable to synthesis, we have overcome the stumbling block provided by the β-ketothiolase reaction of equation (6). Once again, reversal of equilibrium has been bought at the price of a molecule of ATP. The acetoacetyl-

CoA is now free to react with another molecule of malonyl-CoA, thus building up large fatty acid chains.

The present position, then, is that, although the cell probably makes some of its fatty acid by simple reversal of oxidation, the larger part of it is produced by a system of enzymes which, although performing formally similar reactions, in fact differ in certain essential respects. In particular, they use NADP instead of NAD as coenzyme, and bypass the difficult thiolysis reaction by means of new intermediate, malonyl-CoA, at the expense of an extra molecule of ATP. All these facts, taken together, have led to the conclusion that the enzymes of fatty acid synthesis are linked together in the cell into one giant multi-enzyme complex, perhaps sited on the membranes of the endoplasmic reticulum, which performs all the synthesis reactions one after the other, taking in acetyl-CoA at one end, and, with the help of malonyl-CoA, $NADPH_2$, and ATP, turning out stearic, palmitic, and oleic acids at the other. The concept of such synthetic assembly-lines is becoming of importance in many different aspects of biochemistry.

TURNING FATTY ACIDS INTO FATS

Fats are converted into fatty acids by hydrolysis under the influence of esterases (page 161). But, once again, though the esterase system is readily reversible, the route adopted in biosynthesis is more complex, and demands an energy supply in the form of ATP and coenzyme A. At least two ways of making triglyceride fats from the fatty acids have been found. In one, the glycerol is first activated in a now familiar manner, by phosphorylation to give α-glycerophosphate (11):

$$
\begin{array}{l}
CH_2OH \\
|\\
CHOH \quad + \quad ATP \quad \rightarrow \\
|\\
CH_2OH
\end{array}
\quad
\begin{array}{l}
CH_2OH \\
|\\
CHOH \quad + \quad ADP \\
|\\
CH_2O\!-\!\boxed{P}
\end{array}
\qquad (11)
$$

The activated glycerol can now react with first one and then a second molecule of palmityl-, stearyl-, or oleyl-CoA (12, 13):

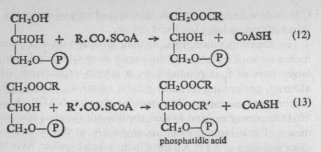

$$\begin{matrix} CH_2OH \\ | \\ CHOH \\ | \\ CH_2O-\text{(P)} \end{matrix} + R.CO.SCoA \rightarrow \begin{matrix} CH_2OOCR \\ | \\ CHOH \\ | \\ CH_2O-\text{(P)} \end{matrix} + CoASH \quad (12)$$

$$\begin{matrix} CH_2OOCR \\ | \\ CHOH \\ | \\ CH_2O-\text{(P)} \end{matrix} + R'.CO.SCoA \rightarrow \begin{matrix} CH_2OOCR \\ | \\ CHOOCR' \\ | \\ CH_2O-\text{(P)} \\ \text{phosphatidic acid} \end{matrix} + CoASH \quad (13)$$

The product is phosphatidic acid, itself an important substance (see page 31) and the starting point for the phosphatides and related lipids. The conversion of phosphatidic acid into triglyceride demands first the removal of the phosphate through hydrolysis by the enzyme *phosphatidic acid phosphatase* (an irreversible reaction). The product is diglyceride, which can react with a final fatty acyl-CoA to yield triglyceride (14, 15):

$$\begin{matrix} CH_2OOCR \\ | \\ CHOOCR' \\ | \\ CH_2O-\text{(P)} \\ \text{phosphatidic acid} \end{matrix} + H_2O \rightarrow \begin{matrix} CH_2OOCR \\ | \\ CHOOCR' \\ | \\ CH_2OH \\ \text{diglyceride} \end{matrix} + \text{(P)} \quad (14)$$

$$\begin{matrix} CH_2OOCR \\ | \\ CHOOCR' \\ | \\ CH_2OH \\ \text{diglyceride} \end{matrix} + R''.CO.SCoA \rightarrow \begin{matrix} CH_2OOCR \\ | \\ CHOOCR' \\ | \\ CH_2OOCR'' \\ \text{triglyceride} \end{matrix} + CoASH \quad (15)$$

Four molecules of ATP are consumed in making fats by this route; one for making glycerophosphate and the other three for the CoA derivatives of the fatty acids. This reaction path is the one followed in adipose tissue where the fat deposits are laid down for the body as a whole. But another slightly more economical route occurs within such cells as those of the intestinal wall, kidney, and liver, where the monoglyceride monolein can react directly first with one and then with another fatty acyl-CoA molecule to give a triglyceride without the need for the prior phosphorylation of glycerol. The enzyme responsible for these reactions is *mono-*

glyceride transacylase, which thus makes triglyceride using only three ATP molecules.

But even with this marginal saving in the efficiency of esterification, the whole process of fatty acid synthesis by the cell is an object lesson demonstrating how much harder it is to build than to destroy, and how much more effort is required. And we shall find this lesson repeated with far more emphasis as we turn from the manufacture of such relatively simple substances as glycogen and fats to the building operations necessary to create proteins or nucleic acids.

CHAPTER 10

SYNTHESIS OF PROTEINS AND NUCLEIC ACIDS

THE UNIQUENESS OF PROTEIN SYNTHESIS

Protein synthesis is one of the most important of the cell's synthetic activities. The growing animal makes protein at an extremely rapid rate; up to seventy per cent of the liver of a growing rat can be removed, and within twelve days the organ will have regained its original weight. But even after the body has achieved its maximum size, protein synthesis still goes on. Like all the giant molecules, proteins are part of the 'dynamic state of body constituents'. Enzymes, haemoglobin, structural proteins such as collagen, are all constantly being broken down and resynthesized at varying speeds. Under normal conditions, the average cell is probably synthesizing several thousand new protein molecules every minute.

The problem is almost different in kind from the problems of the synthesis of those substances we considered in the previous chapter. The fact that the last few years have gone a great way towards solving it represents one of the greatest triumphs of modern biochemistry, as important as the splitting of the atom in physics or the theory of relativity in cosmology. Its solution depended on the inflow of ideas and methods not only of classical biochemistry but also of genetics and the new science of 'information theory'. That its unravelling depended not on the brilliant insight of one individual, but on the hard work over many years of a large number of laboratories in different parts of the world, makes the saga of protein synthesis representative of a new class of scientific discovery – one that depends on the highly organized state of modern science, so that workers in different disciplines and different continents were kept constantly in touch with one another's experiments and thoughts not only by special journals set up to cope with the flow of results, but by letters, teleprinters, conferences – even the daily newspapers.

Why is protein synthesis so different from that of polysaccharides or lipids? The actual chemistry of the formation of peptide bonds between amino acids is of the type of reaction that we are already quite familiar with. Two amino acids can link together with the removal of water to form a dipeptide:

$$NH_2.CH.C\underset{O}{\overset{R}{|}} + N.CH.COOH \rightarrow NH_2.CH.C\overset{R}{\underset{O}{|}}-N.CH.COOH +$$

This reaction is the reverse of those carried out by the protein-splitting enzymes that hydrolyse proteins during digestion, such as trypsin and pepsin. But it cannot be carried out by them, because, like other synthetic reactions, it is energy-requiring, and we may anticipate that it will be necessary to provide this energy in the form of ATP. It is possible to test this hypothesis by studying the synthesis of some of the simple peptides that the cell regularly makes.

One of these is the tripeptide glutathione, γ-glutamylcysteinylglycine, a substance found in most cells although its actual function is not clear. Glutathione is made by the cell in two stages: in the first glutamine and cysteine condense to give γ-glutamylcysteine, and in the second the dipeptide reacts with glycine to give γ-glutamylcysteinylglycine. As expected, two molecules of ATP are split during this synthesis.

But the problem of protein synthesis is more complex than that of providing an enzymic mechanism for peptide bond formation. Glycogen synthesis was solved when we demonstrated an enzyme system which could add glucose on to the growing polysaccharide chain, and a branching enzyme to fork the chain at fixed intervals. But glycogen is made of only one sort of sugar – glucose. Even more complex polysaccharides contain only three or four different sorts of sugar at most, bound together in a repeating pattern. But the protein chain is made of twenty different sorts of amino acid joined head to tail in a unique and non-repetitive sequence. Each amino acid has its predetermined place in the sequence, and if it is altered in some way, by the replacing of one amino acid by another, or by the inverting of the order of two or more along the chain length, it may mean that the protein can no longer fulfil its physiological function within the cell (see, for example, page 42). The protein synthesis problem is one of specificity.

It is now possible to see why both the geneticists and the cyberneticists (information scientists) are interested in protein synthesis. The geneticist is concerned with how cells and organisms reproduce themselves. Classical genetics started in the late nineteenth century with Mendel, who examined the laws by which flowers such as sweet peas crossbred to produce varieties of different colours. Modern genetics, a science which has grown up in the last twenty or thirty years practically contemporaneously with biochemistry, is concerned with the *mechanism* of such reproduction. To the geneticist, the proteins are the molecules that control the day-to-day running of the cell. The proteins of the infant are identical with those of his or her parents. But the infant was produced by the fusion of two microscopic germ cells. These two tiny cells must have had packed within them detailed instructions and blue-prints for the manufacture of all the several thousand different proteins in the billions of different cells of the infant's body.

The cyberneticists, who represent an even newer science than either biochemistry or genetics (the very name was only coined in 1948), come in here. Their job is the study of *information transfer*: the converting of information from one form to another – the human voice into radio waves and back into sound once more, or a complex mathematical equation into a set of punched holes on a tape, to be fed into a computer and then into a set of traces on reels of magnetic tape in the computer's 'memory store'. The theory of transfers of this sort is a cyberneticist's bread and butter. To them, protein synthesis is just such another case. The mechanism for ensuring the exact replication of a protein chain by a new cell is that of transferring the *information* about the protein structure from the parent to the daughter cell, and subsequently of translating that stored information back into the synthesis of a protein chain again.

Let us summarize the situation as it appeared to the biochemists, geneticists, and cyberneticists by about 1950. The problem was how to assemble a series of up to 300 amino acids head to tail along a peptide chain so that the order of the amino acids along the chain

could be accurately predetermined and reproduced. It is possible
to visualize two ways of building up such a chain. Either we could
start with one amino acid, add a second to it, then a third, a fourth,
and so on, until the chain is complete; or, alternatively, we could
collect together all the amino acids needed for the chain, and,
when they were all assembled, link them almost simultaneously
one to another (like doing up a zip-fastener). The second alternative seems an unlikely one, and would not be considered at all
were it not that the first leads into a morass of difficulties. We can
illustrate these difficulties by a hypothetical example. If we wish
to join glycine to alanine, we can envisage an enzyme which performs the reaction:

$$gly + ala \rightarrow gly\text{-}ala$$

(like the glutathione synthesis we described earlier). We can now
imagine a second enzyme to add a third amino acid – say tyrosine:

$$gly\text{-}ala + tyr \rightarrow gly\text{-}ala\text{-}tyr$$

To add a fourth amino acid we need a third enzyme, and to add a
fifth, a fourth enzyme is required. In general, to build a chain of
n amino acids, we need $(n - 1)$ enzymes. But as the protein chain
grows in length, so the amount of information that the nth enzyme
needs in order to add the $(n + 1)$th amino acid also increases. In
deciding whether to add serine as the twenty-fifth amino acid, the
enzyme must have at its active centre a means of checking all the
preceding twenty-four acids to ensure that they are present in the
right order. Now enzymes may be good at the job, but this must
obviously impose an enormous strain upon them. By the time the
299th enzyme adds the 300th amino acid, error is bound to have
crept in. And errors, as we know, are likely to prove fatal. But there
is a still greater absurdity, for we have postulated a process whereby, to make a single protein chain of 300 amino acids, 299 enzymes
are required. Now there are perhaps a thousand different proteins
present in the cell, and to make them we need a total of 299,000
specific enzymes! And to make each of these 299,000 enzymes,
each of which is a protein, we need another 299 enzymes . . .
Clearly the argument is a nonsense.

Experimental evidence backs up theory in ruling out this stepwise enzymic synthesis of protein. If protein were made by a

steady build-up of peptide sub-units, study of a cell that was engaged in active protein synthesis should reveal the presence of many of such peptide fragments. Yet few have ever been found, and those that are found are always specific molecules with particular biochemical roles themselves – like glutathione, for instance – and not merely precursors of completed protein chains. Similarly, if a tissue is fed all but one of the amino acids essential for synthesis, it would be expected that a large number of peptide chainlets would be formed, representing growing protein chains which had been broken off where the missing amino acid should have fitted in. But in fact under such conditions what happens is that synthesis stops completely, and no fragments are ever found. Thus the hypothesis that proteins are formed from amino acids by stepwise enzymic synthesis must be abandoned.

Templates and genes

Yet the alternative hypothesis, as it stands, is equally unsatisfactory. It is difficult to picture a system in which 300 or so free amino acids come spontaneously together in the correct order to link up to form a chain. The odds against it would be immense. Some method must exist whereby the amino acids are assembled at particular sites, held in waiting until all are present and correct, shuffled into an approved order, and then linked. This implies the existence of 'sites' at which the amino acids can be 'parked' pending the completion of the chain. Each site must be able to distinguish between the different amino acids so that only an acceptable one can be fitted in, and the sites themselves must be linked to each other in the same order as the amino acids of the final protein chain (Figure 24, p. 204). In other words, protein synthesis demands a pre-existing set of 'amino acid recognition sites' forming part of a pre-existing macromolecule.

Such an answer to the riddle satisfied biochemists, geneticists, and cyberneticists alike. A 'mould' or 'template' for protein synthesis was predicted by the biochemical requirements, whilst, if it could be shown that the template was under genetic control, the clue to how information on protein structure is transferred from cell to cell would be found. What would be the requirements of a template? It must be a macromolecule related in length to the protein it is synthesizing. Along its length must be a series of sites corresponding to the individual amino acids of the protein, and each such site must be capable of distinguishing between different amino acids and accepting only one. We know of only three types of macromolecular chain: carbohydrates, proteins, and nucleic acids. Carbohydrates can be ruled out because chains, say, of glucose in glycogen cannot possess the high degree of specificity demanded of a structure designed to distinguish between amino acids. Nor can we accept that protein chains are made on other proteins as templates. Although, evidently, a protein fulfils the criteria of specificity we have laid down, it would mean that proteins were a self-reproducing set of molecules. This gives the same sort of logical absurdity that we found when we considered the possibility of a stepwise enzymic synthesis.

There remain the nucleic acids as possible templates. DNA and RNA are macromolecular chains often many times greater in length than proteins. They are composed of a set of four different bases arranged in a complex but ordered pattern along their chain length. Such bases could easily be organized into a sequence of 'sites' that could discriminate between amino acids, and could thus fulfil the criteria we have laid down for the template. Thus the proposition that the nucleic acids form a template on which proteins are synthesized makes biochemical sense.

It also makes genetic sense. It has been known for many years that the nucleic acids are closely involved in the reproductive process. The geneticist finds that when a cell divides, or two cells fuse in sexual reproduction, the instructions that the new cell needs for controlling its growth and development are contained in a small number (in humans, forty-six) of thin strands of material which appear under the microscope like a handful of twisted streamers after a child's party. Those strands are called chromosomes. When a cell reproduces by division, each chromosome splits into two halves, so that each daughter cell contains the same number of chromosomes as its parent. In sexual reproduction, each parent cell provides half the chromosomes of the offspring. Thus, whether produced sexually or by division, every new cell ends up with the same number of chromosomes as the parent cell.

Each chromosome consists of a number of units strung along its length like beads on a necklace. Each unit, or bead, is responsible for the transmission to the new cell of the instructions for the expression of a particular characteristic property which, as the cell develops, will unfold in interaction with its environment. Characteristics such as colour of hair or eyes, or for normal or abnormal haemoglobin, are all examples of the types of property possessed by an organism (its phenotype, as geneticists term it), for which the basic genetic instruction may be found within the chromosome. The beads, or genetic units, are called genes by the geneticists, who deduced their existence theoretically long before the electron microscope enabled the chromosome to be examined visually in detail and the genetic units along its length mapped by a set of genetic techniques such as the use of controlled mutation in bacteria that are in some ways reminiscent of those described for the study of metabolism in Chapter 6.

But in the biochemical picture of the cell that we have built up, the day-to-day running of the cell is the job, not of genes, but of specific proteins; in particular, enzymes. Eye or hair pigments, for instance, are actually made by a series of enzymes, even though the *instructions* about how to make them are contained in the genes. The geneticists therefore argued that the role of the genes was as organizers and controllers of protein synthesis. And when the chromosomal material was found to consist of one of the two nucleic acids, DNA (as was mentioned in Chapter 4), the link-up seemed complete. It is DNA that carries the complex set of instructions from parent to daughter cell. The chromosome is made up of a long, thin, helical DNA molecule, made by the intertwining of two DNA chains according to the Watson-Crick model (Figure 5, page 73) and bound round and protected by a basic protein, histone. The histones are arranged in units of eight molecules (octomers), together with a length of DNA about 200 nucleotides in length, so that under the electron microscope the chromosomes have a beaded appearance. Each of these beads (composed of a histone octomer plus its associated DNA) is known as a nucleosome and probably several of these nucleosomes together represent the basic genetic unit or gene.

Is DNA, then, itself the 'mould' or 'template' for protein synthesis for which we are searching? Several pieces of evidence make this unlikely. Virtually all the DNA of the cell is in the chromosomes, which are themselves, at least in animal and plant cells, contained within the nucleus. In certain organisms, such as the green plant-like algae, it is possible to remove the nucleus, and hence all the DNA, by dissection. The *enucleate* cells of one alga, called acetabularia, can continue to live for some weeks after the operation has been performed, and during this time will continue to synthesize proteins quite as efficiently as the whole cell. Similarly, it is possible to make preparations from cell homogenates which are free of DNA, and which nonetheless will incorporate radioactive amino acids into new protein. Even treating such homogenates with an enzyme (*DNA-ase*) which will remove any last traces of DNA that might be present by hydrolysing it, will not prevent protein synthesis. So we must conclude that, although DNA is the fundamental genetic material, and hence is ultimately responsible for protein synthesis, it does not form the immediate template on which the protein is made.

However, in order to understand how precise copies of proteins can be made in daughter cells, it is necessary also to see how DNA itself is synthesized, for in cell division there must be a foolproof way of copying off the DNA instructions from the parent cell and transmitting it to the daughters.

DNA REPLICATION

Although earlier we argued that, so far as proteins were concerned, one protein could not act as a mould or template for another, this argument does not apply to DNA. DNA synthesis, it turns out, is *not* the synthesis of a molecule from its constituents like that of glycogen or the fats, but the direct 'copying-off' of one DNA molecule from another, a sort of photocopying process known as *replication*. It is the unique structure of DNA which makes this photocopying possible, a point immediately apparent to Watson and Crick when they first 'solved' the DNA structure in 1953 (see page 72).

DNA, it will be recalled (and it is worth looking back at the figure on page 73) is a double stranded molecule composed of two chains of nucleotides held together by hydrogen bonding between the purine and pyrimidine base pairs. During replication, the hydrogen bonds are broken and the two chains separate and unwind. Each then acts as a template on which a new DNA chain can be synthesized from the constituent nucleotides, which become attached in the requisite order because there are no alternative options open for them except for A to pair with T and G with C.

Thus replication is *semi-conservative*, each new DNA molecule containing one of the parent chains plus one new chain. The enzyme responsible for synthesis of the new chain is DNA polymerase, which uses as its substrate the deoxyribonucleotide triphosphates. In fact, synthesis does not begin at one end of the DNA chain and continue until it reaches the other because the DNA molecules are so long that unravelling of the helix from one end continuously would set up an intolerable strain on the rest of the molecule long before the task was completed. Instead synthesis takes place simultaneously in different parts of the molecule, and to make it easier for the polymerase to recognize where it should begin replication the DNA helix has a number of *initiation points* along its length. So the polymerase not only needs to possess binding sites for its four different substrates (ATP, GTP, CTP, and TTP), it also has to have the capacity to recognize and bind to the initiation point and it has to recognize the base on the parent strand so that it can insert the correct matching base into the growing daughter strand. It is therefore not surprising to find that DNA polymerase is yet another example of a large allosteric enzyme.

Yet another problem facing the enzyme is due to the fact that the two DNA strands in the double helix run in opposite directions – in one chain the nucleotides are linked by a phosphate group through the hydroxyl at carbon atom No. 5 (C5) of a ribose sugar molecule to the hydroxyl on C3 of the next sugar in line; in the other chain the linkage is from C3 $\rightarrow$ C5. The polymerase can only synthesize new chains in the $5\rightarrow3$ direction (p.71). What happens is that the enzyme copies one side only at first in the $5 \rightarrow 3$ direction, leaving one strand unpaired until there is enough room to enable another polymerase molecule to initiate replication of the unpaired chain in the correct chemical direction. This explanation of DNA synthesis was first proposed by Okazaki in 1968.

Only one problem remains – it turns out that even at the initiation points the polymerase will not begin *de novo* synthesis unless there is a primer attached to the chain – a very short length of single stranded nucleic acid to which the incoming nucleotide can be attached. It seems now as though this primer may be not DNA at all, but RNA which can later be chopped out from the completed chain and the gap it leaves filled, although the mechanism by which this is achieved is still unknown.

Of course the whole process of DNA synthesis requires a considerable input of energy. We have already said that the substrates for the polymerase are the nucleotide triphosphates, but it is the *monophosphates* that are inserted into the chain. So once again the splitting off of the two terminal phosphate groups from the nucleotides provides the driving force for a biosynthetic reaction.

Hence we have an elegant biochemical method for the copying of the genetic material, a model which economically and precisely accounts for the molecular events underlying genetic transmission of information across generations, and which has become the focus of research attention of molecular geneticists over the 1960s and 1970s. To discuss in any more detail the mechanisms of DNA replication, their association with the events of cell division and with the genetic phenomena such as dominance, recombination, and mutation would take us into new terrain outside the scope of this book. So we leave the DNA story at this point, content merely to see that the process of replication we have described avoids the trap of an infinite regress of enzymes to fabricate a macromolecule with a defined structure with accuracy and efficiency.

RNA AND PROTEIN SYNTHESIS

We have described how DNA is copied, but also made it clear that it is not itself the template for protein synthesis. This function is reserved for the other of the two nucleic acids, RNA. RNA, it will be recalled from Chapter 3, is like DNA, a chain composed of purine and pyrimidine bases, in this case adenine, guanine, cytosine, and uracil. Unlike DNA, the sugar component is not deoxyribose but ribose itself. And, also unlike DNA, its molecules are not composed of two strands intertwined into a helix, but of single chains. There are three forms in which RNA is found in the cell. Some of it is in association with DNA in the nucleus – the so-called *messenger RNA*. Some of it, of relatively low molecular weight, is in solution, and is called *transfer-RNA*. Finally, the bulk exists, as was described in Chapter 4, as tiny insoluble particles, distributed through the bulk of the cell and known as *ribosomes*, which contain about ninety per cent of all the cell's RNA. In the ribosomes, the RNA is found bound tightly to protein from which it can be separated only with the greatest difficulty.

Experiments show that it is on the ribosomes that protein synthesis actually occurs. If cells synthesizing protein from radioactive amino acids are studied, the radioactivity is first found bound to the ribosomes, and is only later released from them as soluble protein. A cell suspension from which ribosomes have been removed can never be made to synthesize protein, whilst if they are subsequently replaced synthesis can proceed rapidly. But the ribosomes alone are inadequate. In order to incorporate radioactive amino acids into new protein there needs to be added to the ribosomes a preparation of soluble cell material which contains certain enzymes, the soluble low-molecular-weight messenger RNA (m-RNA), transfer RNA (t-RNA), ATP, GTP, and ions like magnesium and potassium. So what role do these various substances perform in protein synthesis?

The first task is somehow to transfer the information from the DNA molecule contained within the nucleus to the site where the protein is to be assembled, that is, to the ribosomes attached to the endoplasmic reticulum. This is the job of m-RNA which is synthesized from DNA in a manner very similar to the synthesis of DNA itself, in a process called transcription. Transcription is controlled by RNA polymerase and because the DNA contains instructions for the manufacture of all the cell's proteins, only a few of which are likely to be needed at any one time, the enzyme must be able to recognize or select which particular DNA regions, representing the desired proteins, need to be transcribed at any one time. It probably does this by attaching to an area of DNA characterized by having a specific nucleotide sequence – this is the so-called *promotor site* and it is not until the enzyme is actually bound to the DNA that the chain opens to reveal the template for transcription. Chain opening most likely involves a conformational change in the DNA molecule that forces the hydrogen bonds between the bases apart. The enzyme can now synthesize an RNA strand in the 5 → 3 direction by base pairing, just as when DNA is being copied. Transcription differs from DNA replication in that only one of the strands is copied at any one time, although which strand this is depends on which protein is to be made. It must also be remembered that because of the mechanism of base pairing the RNA will be an exact copy not of the template DNA but of the *opposing* DNA strand. Another difference is that because RNA contains uracil instead of thymine, it is this uracil that pairs with the DNA adenine base. The process is shown diagrammatically in Figure 21.

This mechanism enables a length of RNA to be produced as a copy of the genetic material, the DNA, which corresponds to the protein which is ultimately required. However, not only must the RNA polymerase have a means of recognizing where to *start* synthesizing RNA on the DNA chain, it must also know where to stop when the appropriate length of DNA has been transcribed. The factors which ensure that RNA synthesis stops are not fully understood; a protein described simply as the rho factor is involved. The length of newly synthesized m-RNA is released from the DNA chain and the DNA double helix re-forms.

The m-RNA must now leave the nucleus for the cytoplasm. It does this, it is believed, by way of the pores which are visible in the nuclear membrane (Chapter 4, and Plate 1). In the cytoplasm the m-RNA becomes slightly modified in such a way as to enable it to recognize, and bind to, the ribosome. There we must leave it for a moment, and return to the question of how the amino acids to be turned into protein are prepared in their turn for attachment to the m-RNA template.

The first step in this preparation consists of the activation of the amino acids, so as to prime them with sufficient energy for the subsequent formation of peptide bonds. This activation is performed at the expense of ATP by a series of amino acyl synthetase enzymes, one for each amino acid. These enzymes were reported

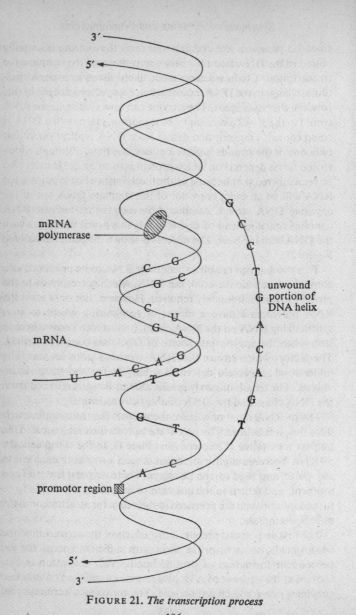

3′

5′

mRNA
polymerase

unwound
portion of
DNA helix

G
C
C
C
T
G
C
A
C
A
G
T
T
T
C

C
G
G
C
C
U
G
A

mRNA

U G A C A
T
C
G

promotor region

5′

3′

FIGURE 21. *The transcription process*

196

first by M. Hoagland in America, who observed that as a group they were all precipitated from a soluble cell preparation by acidifying to pH 5.0. They are thus sometimes known as 'pH 5 enzymes'. In their presence, amino acids react with ATP to form amino acid-AMP complexes:

$$H_2N.\overset{\overset{\displaystyle R}{\displaystyle |}}{CH}.COOH \ + \ ATP \ + \ ENZ \ \rightarrow$$

$$ENZ.-AMP.\overset{\overset{\displaystyle R}{\displaystyle |}}{CO}.CH_2NH_2 \ + \ pyrophosphate$$

In the next stage of preparing the amino acid for the template the activated amino acid is transferred from the enzyme complex to a transfer-RNA (t-RNA) molecule, of which there seem to be about two or three alternative ones for each amino acid. Many t-RNAs have been purified, and their structures determined – indeed t-RNAs, which are around eighty nucleotides in length, were the first nucleic acids to be sequenced. The synthetase enzyme catalyses the formation of a covalent bond between the carboxyl group of the amino acid and the free hydroxyl group at the C3 terminal end of the t-RNA molecule, releasing the synthetase enzyme once more. All t-RNA molecules have been found to have the same trinucleotide sequence of cytosine–cytosine–adenine (CCA) at the end which links to the amino acid, and they all seem to be folded into the same secondary structure, rather like a four-leaved clover (Figure 22). The molecule thus has three other arms besides the one that becomes charged with the amino acid. These other arms of the clover leaf are characterized by the presence of certain unusual nucleotides, and each has a specific function. One of the arms, for example, is responsible for recognizing the ribosome to which the t-RNA is shortly to become attached.

To recapitulate the story to date, we have arrived at a situation where each amino acid about to become incorporated into protein is now linked to a specific t-RNA. To arrive at this point, the cell has already had to provide more than twenty different enzymes, one for each amino acid, and twenty different t-RNAs, as well as two phosphate bonds per activated amino acid. It has also been necessary not only that the enzymes be specific for individual

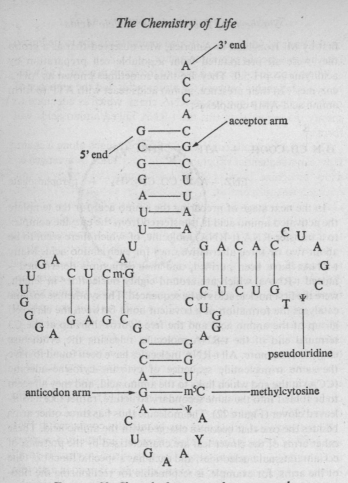

FIGURE 22. *Clover leaf structure of yeast t-RNA^phe*

amino acids, but that a new class of specific molecules is introduced – a set of t-RNA molecules each capable of 'recognizing' a different amino acid. Meanwhile, we have established that, attached to the ribosome, is a strand of m-RNA which in some way contains a copy of the instructions, originally present in DNA, on the basis of which the activated amino acids are to be linked together in a specific protein sequence.

We now come to the key aspect of the entire process. There has to be a relationship between the amino acids and the sequence of nucleotide bases along the m-RNA chain, which is adequate to specify the sequence in this the t-RNA linked amino acids will form up.

The RNA molecule consists of 4 bases arranged along a chain. How can a sequential structure of 4 different units be arranged so as to 'recognize' a set of 20 different units? The answer to this question was provided in the mid 1950s by the physicist George Gamow. He argued that, if each base corresponded to a different amino acid, only 4 different amino acids could be recognized by the chain of bases. If it needed a combination of 2 bases to code for each amino acid, 16 different amino acids could be recognized, as there are 16 (or 4×4) different possible ways of combining the 4 different types of base into groups of 2. Similarly, if 3 bases corresponded to each amino acid, 64 (or $4 \times 4 \times 4$) different amino acids could be recognized (Figure 23, p. 202). As there are up to 20 different amino acids found in proteins, the minimum number of bases that could correspond to each 'recognition site' along the RNA chain must be 3. Gamow therefore suggested that, along the RNA template, bases were organized into successive groups of 3, each group corresponding to a particular amino acid. An RNA chain of 600 bases would therefore be needed to code for a protein chain of 200 amino acids.

This theoretical prediction was beautifully confirmed in 1961 and 1962 by Francis Crick, Sidney Brenner, and their co-workers in Cambridge, in genetic studies with a virus. A virus is to all intents and purposes a packet of nucleic acid wrapped in a protein jacket. It exists by attacking cells – animal, plant, or bacterial – taking over the hosts' protein and nucleic acid synthesizing systems and using them to make more *viral* protein and nucleic acid. With these it replicates its own structure several times, until it has used up and exhausted all the utilizable substrates present in the host cell. Finally, the virus bursts the host cell, releasing several new virus particles to hunt for fresh prey (see Plate 8). Viruses that attack bacteria are called *phages*, and, with one of these, Crick and Brenner were able to show that the correspondence:

$$3 \text{ bases} \equiv 1 \text{ amino acid}$$

did indeed hold.

It became apparent during the 1960s that this recognition mechanism was not confined to viruses or bacteria but was the universal code for the synthesis of proteins throughout the living world.

The m-RNA strand is thus arranged in a series of units, each three bases long, each of which represents, or codes for, a particular amino acid. Corresponding to these three m-RNA bases are a complementary set of three bases on the arm of the t-RNA molecule directly opposite the arm which binds the amino acid, and these three nucleotides form base-pairs with the m-RNA triplet on the ribosome. Each triplet of three bases along the m-RNA is termed a codon; the complementary triplet on the t-RNA is called an anti-codon.

We thus have a protein synthesizing system in which suitably activated amino acids are brought into contact with an RNA chain where each individual amino acid is 'recognized' by a group of three bases and held in position by them. When enough of the amino acids required to match all the groups of bases on the RNA chain are present, the peptide bonds between adjacent amino acids can be formed, the activating t-RNA-molecule released into solution, and the new protein molecule unpeeled from its RNA template.

Is there any way of telling which triplets of bases correspond to which amino acids? The answer to this question was brilliantly provided by Marshall Nirenberg and Heinrich Matthaei in Maryland in 1961. Nirenberg and Matthaei argued that if a synthetic RNA chain made up of only one type of base could be made, such a chain ought to provide a template for the synthesis of peptide chains containing only one type of amino acid. By using an RNA synthesizing enzyme, *polynucleotide phosphorylase*, they were able to make an artificial RNA chain containing only the base uracil (*polyuridylic acid*). When they tried the effect of this in a protein synthesizing system, in which the artificial RNA chain replaced the natural m-RNA, they discovered that in its presence a peptide chain containing only one sort of amino acid – phenylalanine – was made. Thus the sequence of bases U-U-U is the RNA recognition site for the amino acid phenylalanine. This breakthrough was rapidly followed by both Nirenberg and Matthaei and by Severo Ochoa in New York. RNA polymers made up of differing combinations of the four bases yielded differing artificial peptide chains, and by the end of 1966 virtually the entire RNA code language for amino acids was known. Polynucleotide phosphorylase was indeed the Rosetta stone of protein synthesis.

The relationships between nucleotide triplet 'codons' and the amino acids for which they code is shown in Figure 23. The sixty-four possible combinations of the 4 RNA bases into groups of three codes for only twenty amino acids, and there is therefore some redundancy amongst the codons; three different groups, U-C-U, U-C-C and U-C-G all code for the same amino acid, serine, for instance. In the parlance of the cyberneticists, the RNA code for proteins is 'degenerate'. Several of the codons, though, do not represent amino acids at all, but are signals which mean, for instance, 'start reading' and 'stop reading' – that is, they indicate the beginning and end of a protein chain, just as a capital letter and a full stop imply the beginning and end of a sentence. The 'full-stop' codons are U-A-G, U-A-A and U-G-A, whilst the 'capital letter' system is rather more complicated, as we will see in a moment.

Let us look more closely at the molecular events at the ribosome, where the recognition and formation of peptide bonds actually occurs. We have seen that the ribosome is composed of two different subunits, but only one of these subunits, the smaller of the two, is essential for initiation of protein synthesis, although it must be associated with the larger unit before chain elongation can proceed. Initiation also requires the presence of an energy source (supplied not by ATP but GTP), a particular amino acyl t-RNA whose anti-codon corresponds to the 'start here' codon on m-RNA and, at least in bacteria, three soluble protein 'initiation factors' called IF1, IF2, IF3. The ribosome has two sites for t-RNA binding, the P site and the A site, but only initiator t-RNA can bind to the P site – all other incoming amino acyl t-RNAs bind to the A site.

FIGURE 23. *The Genetic Code*

1st↓ 2nd→	U	C	A	G	↓3rd
U	PHE	SER	TYR	CYS	U
	PHE	SER	TYR	CYS	C
	LEU	SER	—	—	A
	LEU	SER	—	TRP	G
C	LEU	PRO	HIS	ARG	U
	LEU	PRO	HIS	ARG	C
	LEU	PRO	GLUN	ARG	A
	LEU	PRO	GLUN	ARG	G
A	ILEU	THR	ASPN	SER	U
	ILEU	THR	ASPN	SER	C
	ILEU	THR	LYS	ARG	A
	MET	THR	LYS	ARG	G
G	VAL	ALA	ASP	GLY	U
	VAL	ALA	ASP	GLY	C
	VAL	ALA	GLU	GLY	A
	VAL	ALA	GLU	GLY	G

The protein chain grows in a particular direction, so that the first amino acid of the sequence has a free NH_2 group whilst the last amino acid has a free COOH. Experiments have shown that in most cases the N-terminal amino acid is methionine. So Met-t-RNA would seem to be the initiator. But methionine can also be inserted at other positions in the chain, so how does Met-t-RNA know where to place its amino acid? It is not codon directed, as there is, it is apparent from Figure 23, only one trinucleotide sequence corresponding to methionine, the AUG triplet. In bacteria at least (in which most of these experiments have been carried out), it seems that to serve as an initiator, the methionine must be converted to the unusual amino acid formyl-methionine for attachment to the P site. It seems probable that a similar mechanism exists in mammalian cells although the position here is not yet completely clear.

So once the formyl-methionine initiator is in place, attached to the m-RNA on the A site of the ribosome, chain elongation can take place. The m-RNA is read off as a *non-overlapping* triplet sequence so that the sequence of bases

AUGUUUCAGACC

would be read

AUG	UUU	CAG	ACC
Met	Phe	Glun	Thr

The binding of t-RNA to the next m-RNA codon, on the P site, can then occur (at the expense of GTP). The peptide bond between the amino acids side by side on their respective t-RNA chains can now be formed, catalysed by the binding of t-RNA corresponding to the next codon situated on m-RNA in the P site. It is to supply the energy for this binding that GTP is needed. The next step is peptide bond formation catalysed by the enzyme peptidyl trans-ferase. This leaves a dipeptide attached to the t-RNA on the A site and frees the first t-RNA. As amino acyl t-RNAs can only bind to the A site, before further chain elongation can occur, the whole complex of m-RNA with its attached dipeptide t-RNA must be moved along to the P site, bringing the next m-RNA codon to the A site. This process (translocation) again requires GTP and the participation of a protein elongation factor EF-1. Now the whole process can repeat itself, until, theoretically, the last trinucleotide sequence on the m-RNA chain is decoded. The mechanism is shown diagrammatically in Figure 24. It can best be compared to the functioning of a tape recorder in which the ribosome acts as the pick-up head moving steadily past the m-RNA tape, reading the RNA code as it goes and converting the code words into the language of proteins. Like a tape recorder, the ribosome is neutral as to the message it plays; this is provided by the m-RNA tape. The ribosomal machinery is available for whatever m-RNA tape (or even artificial RNA, as in the Nirenberg and Matthaei experi-ments) appears on the scene. The exact role of the many different ribosomal protein and RNA components in ribosomal function is the subject of active research at present but is still far from clear.

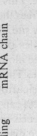

FIGURE 24. *Protein synthesis on the ribosome*

(a) Growing polypeptide chain, attached by the terminal tRNA group to protein-binding site

(b) Attachment of specific AA-tRNA molecule by hydrogen bonding to $(n+1)$ codon of mRNA chain

(c) Formation of peptide bond between AA_3 and AA_4. Ejection of tRNA$_3$

(d) Movement of growing polypeptide chain from AA-tRNA binding site to protein-binding site. Simultaneous movement of mRNA to place $(n+2)$ codon at the AA-tRNA binding site

204

Obviously the physical stability of the ribosome is of prime importance to protein synthesis and it is in the maintenance of this stability that the Mg^{++} and K^+ ions play a part by ensuring the correct ionic environment to prevent dissociation of the ribosomal structure.

Although we have said that translation of the codons could continue to the end of the m-RNA this does not usually happen. Instead translation is terminated by codons that do not bind to amino acyl t-RNA; that is, they do not correspond to a particular amino acid, the full-stop codons referred to above. Release of the polypeptide is probably also mediated by a peptidyl transferase, but one which uses water rather than another amino acid as the peptidyl acceptor. GTP and yet another protein factor are also required for release. After chain release m-RNA and t-RNA remain bound to the ribosomes, but we know that the ribosomes can be used several times, so there must be an as yet undiscovered mechanism for releasing the RNA and regenerating free ribosomes.

This read-out system makes it possible for more than one portion of the long template molecule to be read, by different ribosomes, at a given time and, in fact, in cells which are actively synthesizing proteins, it is possible to isolate, by suitably gentle techniques, such template RNA molecules with several ribosomes attached to different portions of them. Such a cluster of ribosomes held together by a template is called a 'polysome'. This simultaneous read-out enables many molecules of the same protein to be made simultaneously in response to cellular demand.

It is apparent from this that each m-RNA molecule can be used to read out a protein several times over (just as a tape can be played many times on a tape recorder). How long the m-RNA lasts varies from minutes to days, depending on the cell and protein concerned. Some types of template, at least, must be capable of making several hundred protein molecules before needing renewal, as experiments demonstrating protein synthesis in the absence of nuclear DNA reveal.

Not all the details of the protein synthetic mechanism are known (how far mechanisms studied in bacteria can be extrapolated to more complex organisms is just one example of the remaining

problems) but the period since the Watson-Crick paper of 1953 has
undoubtedly proved the most intellectually glorious in the history
of biology in the extent to which what Crick has called the 'central
dogma' of molecular biology, the sequence DNA → RNA →
protein, has become understood. New and important details are
still being added, almost daily. There are indeed molecular biologists who would claim that the 'golden age' is now past, that
most of the vital issues are now resolved, and there is nothing left
to do but fill in the odd pieces of the jigsaw. Further, many molecular biologists believe – and are very articulate in defending the
claim that – not merely all of biochemistry, but of physiology and
even ecology and neurobiology, can be 'reduced' to the inexorable working out of the genetic programme by way of the central
dogma.

It is true that, whereas when the first edition of this book was
written it still seemed adequate to analyse cellular properties
largely in terms of energy flow, it is now necessary to consider
another mode of analysis, that of information flow – and this will
become more apparent as we move to the questions of control of
metabolism discussed in the next chapter. Nonetheless, to argue
that all of biochemistry can be subsumed under the unilateral flow
of information DNA → RNA → protein disregards (a) *which*
proteins are expressed at any time from a particular DNA, and
(b) *how* those proteins function in the cell, and how they are modified and modulated by the environment of the cell itself and the
past history of that cell. To give primacy to any one component
is to miss the interactions that bind them all. It is to these that we
should now turn.

CHAPTER 11

CONTROLLING THE CELL

The last chapters have presented the cell to us as a highly organized industrial plant. We have pictured raw materials, such as amino acids, entering through the cell wall and being processed on a production line of ribosomes geared to produce proteins on a blueprint laid down by a nuclear planning office. These proteins are then sent to other parts of the cell to themselves perform further operations on fresh raw material. Meanwhile, the cell's fuel, glucose, is pulped down to pyruvic acid and dispatched to the mitochondrial boiler-house, where it is converted to useful energy which is distributed throughout the cell and used to drive the entire machinery forward. This is a beautiful picture of an elegant piece of interlocking machinery: each part of the cell needs all the others in order to survive. Without ATP, no protein synthesis; without proteins, no ATP.

But something is still missing from this picture. Industrial processes, as we know, however efficiently organized, cannot operate smoothly and without attention indefinitely. Crises and emergencies arise, production lines become fouled, raw materials are erratic in arriving and half-finished goods may accumulate. Changes of plan are frequently required to readjust the factory's output to the needs and realities of a changing situation. All these changes, in response to both internal and external pressures, require that decisions are constantly made, decisions which must be based on information about activities everywhere within the factory, and on the ability to assess fluctuating and transient conditions. And if so comparatively crude a system as a factory needs its decision-making system, one might argue, how much more so must the delicate and sensitive mechanism of the cell? Is it not here, perhaps, that we arrive at a point which we cannot explain in chemical and physical terms but need instead to evoke a supra-chemical property of life itself to describe?

Today's biochemists would have to answer this question, like others in which 'life' as a non-chemical and physical force has been proposed, 'No'. For them, even the control and regulation of the cell, not only in its day-to-day running, but in its entire cycle from birth to death within the body, and indeed the entire cycle from life to death of the body itself, are interpretable in strictly biochemical terms. The explanation of these regulatory processes is very far from complete; indeed, the question has only relatively recently been recognized as one properly within the domain of biochemistry and, like many other advances in biochemistry, it is an explanation that has had to wait on the development, by scientists of other disciplines, of the tools and concepts needed to analyse and understand the problem.

Decision-making in the running of factories may be assisted – and its necessity partly eliminated – by the process which has had coined for it the ugly name of 'automation', part of the vocabulary of the science of cybernetics. And interpretation of control mechanisms in biochemistry has in fact gone hand-in-hand with advances towards the automation of factories. The cell, in the light of these new theories, begins to appear as the most completely automated factory we know. By this we mean that it is becoming possible to understand how the complex regulatory processes that in the past appeared bafflingly to be the product almost of *conscious will* by the cell are in fact no more and no less than the inevitable consequences of the combination of the appropriate organization of physical structures and chemical reaction sequences that occur within the cell, following inexorably from them according to laws which can be given mathematical form. Let us look at some of these laws.

We can begin by examining the factors that regulate a simple enzymic reaction. The reaction

$$X \underset{\text{X-ase}}{\rightleftharpoons} Y \qquad (1)$$

is catalysed by the enzyme X-ase. We can write a long list of conditions which affect the *rate* at which X is converted into Y. They include the *concentrations* of X and Y, for, assuming the reaction is fully reversible, the substance that exists in the highest concentration at any time will force the reaction to proceed in the direction of its conversion to the other. This is the law of mass action. The *amount* of X-ase present will affect the rate too, because if there is just a small amount of enzyme it will soon become saturated with its substrate and will become the sole limiting factor; a further increase in substrate concentration will have no effect on the reaction (page 106). We have also seen how *temperature* can affect the rate of the reaction, even destroying the enzyme completely by denaturation if it is too great. A change in the *ionic concentration* surrounding the enzyme molecule can have a drastic effect on its activity because of the need to maintain the correct electrical charges at the active centre. Many enzymes too, as we have seen, require the presence of various *cofactors* before they can function, and of course enzymes – especially allosteric ones – are very sensitive to the presence of *activators* or *inhibitors* whose binding causes a conformational change that alters the shape of the active site. Finally, the thermodynamic *equilibrium* of the reaction is also important as it may lie in the direction of complete conversion of X to Y, of almost no conversion, or at some point midway between the two. Any or all of these factors will affect the rate of the reaction; that is, the number of molecules of X converted into Y every minute.

Alteration of any one of these variables will alter the rate of the reaction, although in general it is found that change in any particular one of the variables has a much greater effect than changes in the others. For example, the effects of small changes in temperature and pH may be negligible, and all reactants may be present in optimal concentrations, but X-ase may be obligatorily dependent on magnesium as an activator. Under such circumstances, *very slight* alterations in the amount of magnesium present may have *very large* effects on the rate of production of Y.

Now let us look at the variables affecting a reaction pathway, say the conversion of W to Z by way of the intermediates X and Y:

$$W \rightleftharpoons X \rightleftharpoons Y \rightleftharpoons Z \qquad (2)$$

W-ase   X-ase   Y-ase

For each of the three reactions involved, the set of variables we have described will control the reaction rate. But, over the entire reaction sequence, an additional variable will now operate that did not exist when only one reaction was being discussed, and that is the fact that *each reaction is also affected by the outcome of the others*. Suppose, for example, that the W-ase reaction is extremely sensitive to changes in pH, and that the final reaction product, Z, is an acidic substance. As more and more Z is produced, the acidity of the solution will increase. But when this happens, the enzyme W-ase will be affected and the rate of production of X from W slowed. As the amount of X produced declines, so will the amounts of Y and Z. Thus the Z content will cease to rise and the acidity of the solution will decrease. Immediately, the rate of W-ase will once more be accelerated, Z and acidity will rise, and W-ase will decline. Ultimately, a steady state will be reached.

Thus the production of Z exerts a *controlling influence* on W-ase, and, reciprocally, the rate of W-ase controls the production of Z. This type of control is called *feedback*. It is identical in type to the principle of a thermostat on an electric heater, or the governor of James Watt's steam engine nearly 200 years ago. We can represent a reaction sequence with feedback like this

$$W \rightleftharpoons X \rightleftharpoons Y \rightleftharpoons Z (+ H^+) \qquad (3)$$

W-ase   X-ase   Y-ase

Feedback can be of two types. In the type we have considered, Z production *inhibited* W-ase – this is *negative feedback*. But if Z production lowered the pH so that W-ase was accelerated and Z production increased, this would be *positive* feedback. In all feedback reactions, the principle is the same – the rate of a reaction is controlled by a substance which is an ultimate product but not itself directly involved in the reaction.

When we come to consider the thousands of reactions and hundreds of reaction sequences that occur within the cell, all of them controlled by many different variables, and many themselves altering these variables by generating acidity or alkalinity, using or producing cofactors and inhibitors, our immediate feeling may well be one of despair. With so much going on, how can one sort out just what is, and what is not, significant?

But though it certainly remains complex, one can simplify the problem. Let us consider again our reaction sequence:

Controlling the Cell

$$W \rightleftarrows X \rightleftarrows Y \rightleftarrows Z$$

W-ase   X-ase   Y-ase

If we ask what is the *maximum rate*, under the most favourable circumstances, of each of the enzymic reactions involved, and then express these rates in arbitrary units, we may find, say:

$$\begin{aligned} W &\rightarrow X = 100 \\ X &\rightarrow Y = 10 \\ Y &\rightarrow Z = 1 \end{aligned} \qquad (4)$$

We are saying that the W-ase reaction is ten times as fast as the X-ase reaction, and one hundred times as fast as the Y-ase reaction. What is now the *rate-limiting* step in the production of Z from W? Clearly W-ase is producing X at ten times the rate that X-ase can use it, whilst X-ase is also producing Y at ten times the rate that Y-ase can use it. As both X and Y are thus being produced far in excess of the amounts which can be handled by their respective enzymes, the rate of conversion of W to Z is limited only by the rate of the Y-ase reaction $Y \rightleftarrows Z$. *The rate of a reaction sequence is controlled by the rate of the slowest of the individual reactions of that sequence.* Under these circumstances, even if Z production results in a change in pH big enough to cause a tenfold decline in the rate of W-ase, the rate of Z production will remain unaltered, as this depends not on W-ase but on Y-ase. Z will no longer be exerting feedback control over its own production.

Thus the only reaction which matters so far as the control of a reaction sequence is concerned is the *slowest* reaction of the sequence, and the cell can regulate the rate of any given sequence of reactions by altering the variables which control the rate of just one of those reactions – the slowest. This is where a study of enzyme kinetics and a knowledge of parameters such as Vm, as described in Chapter 5, can be helpful. We may thus expect to find in every reaction sequence one or more critical control points and we may also expect to find them at an early reaction in the pathway, as this would prevent the futile use of enzymes further along the pathway and the build-up of useless intermediates. Similarly, we could expect control points to occur very close to the branching point of pathways that serve more than one function, so that one branch could be regulated without affecting the others.

This, at least, is what theory predicts. How far can we use this theory in practice, in searching for the rules that govern the behaviour of the cell?

INTERNAL REGULATORS OF METABOLISM

I. *Control of energy production*

The bulk of the energy demands of the cell are met within the mitochondria by the production of ATP during the oxidation of substrates by way of the hydrogen transport line (see Chapter 7). When the enzymes and carriers of this system are studied in isolation, they are found to be capable of extremely rapid reactions, yet if the intact mitochondrion is presented with substrates such as pyruvic acid it is found that the rate of pyruvic acid oxidation reaches a maximum which is considerably below the maximum velocities shown by the individual carriers. As increasing the amount of pyruvic acid does not alter this oxidation rate, it is clear that the mitochondrion must contain its own built-in control system to limit the rate at which it burns fuel. We can isolate some of the elements in this control system if we draw a schematic flow-sheet of the operations involved in oxidation (Figure 25).

From this diagram we can see that the rate of oxidation of substrate will be held up if:

(a) not enough oxygen is present to cope with the production of reduced cyt. a_3 at the end of the transport chain. But normally, unless poisons such as cyanide or carbon monoxide are present, the cell has plenty of oxygen available and the oxidation of cyt. a_3 does not represent a rate-limiting step; or

(b) not enough inorganic $\textcircled{P}$ and ADP are present to allow the reverse ATP-ase (which is the coupling factor linking oxidation to ATP production) to work.

This may not be immediately obvious, but consider what would happen if either ADP or $\textcircled{P}$ were not present. Energy trapped within the proton gradient across the membrane could not be tapped off for ATP production. The chemiosmotic pump would have to work at an impossibly high rate to make sure of storing all the

FIGURE 25. *Flow-sheet for oxidative phosphorylation*

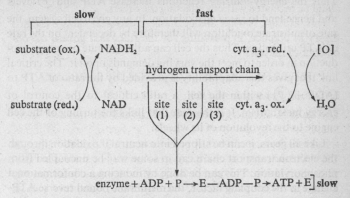

$$\text{enzyme} + \text{ADP} + \text{P} \longrightarrow \text{E}-\text{ADP}-\text{P} \rightarrow \text{ATP} + \text{E} \rbrack \text{slow}$$

ATP synthesis by reverse ATP— ase

energy so the only thing to do, assuming there is plenty of oxidizable substrate, would be to slow down the electron transport chain leading to a build-up of the first reduced hydrogen carrier $NADH_2$. When the levels of ADP and (P) within the mitochondria rise again, oxidation can proceed smoothly and rapidly. It therefore follows that the rate of oxidation of substrate through the electron transport chain is controlled by the concentrations of ADP and inorganic phosphate present in the mitochondrion. When there is plenty of ADP and phosphate, oxidation proceeds smoothly and rapidly. When the concentrations are low, oxidation is correspondingly slow.

But the concentrations of ADP and inorganic phosphate are themselves controlled by other factors. The terminal reaction of oxidative phosphorylation tends to lower their concentration by removing them as ATP. Thus, left to itself, the mitochondrion would gradually remove all the ADP and (P) and cease to oxidize substrate. But at the same time, in other parts of the cell, ATP is being utilized to provide energy for protein synthesis, muscular contraction, nervous transmission, and so forth. All these activities result in the net reaction

$$\text{ATP} \rightarrow \text{ADP} + \text{(P)} \tag{5}$$

Thus the energy-*utilizing* reactions increase ADP and Ⓟ levels and hence tend to *speed up* oxidation. In any given cell system, the rate of substrate oxidation will therefore be dependent on the rate of ATP utilization. Thus the cell can adjust its rate of energy production in order to meet the varying demands upon it. The critical link that gives this adaptability is provided by the ratio of ATP to (ADP + Ⓟ) within the cell, a ratio critical to the control of energy metabolism. It is the gear that links the turning of the cell engine to the revolution of its wheels.

Like all gears, it can be slipped into neutral if oxidation through the electron transport chain can in some way be uncoupled from phosphorylation. This can be done by inducing a conformational change in the coupling factor, the membrane bound reverse ATP-ase. We have seen how the catalytic activity of this enzyme is dependent on the strongly hydrophobic environment provided by the membrane. A conformational change in the enzyme molecule will expose the active site to H_2O on the edge of the membrane, so it can then act as an ATP-ase, hydrolysing ATP to ADP and inorganic phosphate. So temporary uncoupling will decrease the ATP/ADP ratio until it reaches a level when the concentration of ADP will no longer be inhibitory and oxidation can once more be linked to phosphorylation.

Several drugs and antibiotics, and other poisons, possess this uncoupling power. More interesting to the biochemist, perhaps, is the fact that a naturally occurring substance of the body, the hormone *thyroxine*, produced by the thyroid gland in the neck, also seems to be able to uncouple oxidation from phosphorylation, though this is certainly not its primary site of action.

There is always some ATP and ADP present in the cell, and the ratio between them is an important controlling factor not merely within the mitochondrion. Thus in the early stages of glycolysis the enzyme phosphofructokinase, which phosphorylates fructose-6-phosphate to the more reactive fructose 1, 6, diphosphate (page 144) is an allosteric enzyme which is activated by the binding of ADP and inhibited by binding ATP. The need to control both glycolysis and oxidative phosphorylation by the ATP/ADP ratio is not only in the interests of controlling oxidative rates. For example, there may be occasions when the cell still needs rapidly to oxidize food substances to obtain biosynthetic precursor molecules even though ATP levels within the cell are already high and ADP levels are correspondingly very low.

If ADP, inorganic phosphate and oxygen are all present in abundance, none of the control mechanisms we have just outlined will be rate-limiting for oxidative phosphorylation. Under these circumstances yet another reaction becomes the limiting factor, the *entry* of hydrogen into the carrier sequence by way of the dehydrogenase reactions:

$$AH_2 + NAD \rightleftharpoons A + NADH_2 \qquad (6)$$

Like all reversible reactions, the direction in which the dehydrogenase reaction is driven will depend on the relative concentrations of the various products and substrates present. The reaction will thus be driven forward only when the amounts of AH_2 and NAD are large and those of A and $NADH_2$ small. As there are many dehydrogenase reactions supplying hydrogens to the carrier line, all having NAD as a common cofactor, the rate of all of them will depend on the relative amounts of NAD and $NADH_2$ present. When NAD is abundant and the reduced $NADH_2$ only present in low quantities, all the dehydrogenase reactions will tend to move forward, the tricarboxylic acid and fatty acid oxidation cycles that link them will begin to revolve, and a steady stream of hydrogen atoms will be dispatched down the carrier line to oxygen. But if $NADH_2$ begins to accumulate and NAD diminish, the dehydrogenase reaction will slow down to a halt and hydrogen transport cease.

Now, consider what happens if a muscle cell, say, steadily producing energy and oxidizing substrate, is called upon for a sudden burst of activity. ATP levels are reduced, ADP levels rise, and the electron transport chain accelerates. As it accelerates, $NADH_2$ is reoxidized more rapidly and the ratio of NAD to $NADH_2$ increases. As this ratio increases, the dehydrogenase reactions of the oxidation cycles are pushed forward more rapidly and the substrates of oxidation burn yet faster as the cycles spin round. The cell thus provides more energy in response to demand in the same way as the car engine revolves more rapidly at the touch of the accelerator pedal. As the demand for energy slackens, ATP and NADH$_2$ accumulate, hydrogen transport slows down, and the oxidation cycles reduce to a tickover rate only.

There is another interesting side to the controlling role played by the NAD/NADH$_2$ ratio in energy metabolism. We have already referred to the fact that there exist two closely related coenzymes, NAD and NADP, both of which can be alternately oxidized and reduced. NAD is the coenzyme involved in the cell's oxidative, energy-providing systems, whilst NADP finds a place in the synthetic processes, such as fat synthesis, which requires a steady supply of hydrogen for reducing reactions. For these reactions, NADPH$_2$ is demanded, and, just as the ratio of NAD to NADH$_2$ is a control point for oxidation, so the NADP/NADPH$_2$ ratio is a rate-limiting factor for synthetic reactions.

We can now appreciate the rationale behind the specificity of usage of the two co-enzymes because if only one of them was used in both types of reaction it would be impossible to control anabolism and catabolism independently of each other by this method of altering the ratios. An even finer control is exerted by enzymes called *transhydrogenases* which catalyse the reversible reaction

$$NADH_2 + NADP \rightleftharpoons NAD + NADPH_2 \qquad (7)$$

This reaction thus simultaneously oxidizes one coenzyme and reduces the second. By so doing, it serves to alter simultaneously and in reverse directions the ratios NAD/NADH$_2$ and NADP/NADPH$_2$. By altering these two ratios, which between them control the rates of the oxidative and reductive pathways of the cell, the transhydrogenase enzyme helps to hold the balance between the destructive, energy-yielding reactions on the one hand, and the constructive, energy-requiring reactions on the other. And there are interesting suggestions which have been made that the balance point of this see-saw, too, is influenced by hormones, this time the steroid and sex hormones.

II. *Control of biosynthetic pathways*

Discussion of the transhydrogenases has moved us from oxidative to synthetic pathways. Of the many biosynthetic pathways which have been studied in detail in the last half century we have in this book found space to discuss only a few: fat, glycogen, and protein synthesis. We have scarcely referred at all to the means whereby the cell fabricates the precursors of these big molecules – amino acids and purine and pyrimidine bases for example. Yet for the synthesis of each of these there is a chain of enzymes which builds complex from simpler molecules. Each chain will have its own specific rate-limiting steps and feedback mechanisms, many of which have now been analysed in detail, and in particular in organisms which are anyhow more biosynthetically versatile than man – such as bacteria.

Although each biosynthetic system shows certain unique features, there are some general mechanisms that seem to apply. One such was provided by Edwin Umbarger, of New York, in 1956. He was studying the biosynthesis in micro-organisms of the amino acids lysine, methionine, and threonine from glucose by way of a simpler amino acid, aspartate, and its metabolite aspartate-semi-aldehyde. When the bacteria were fed radioactive glucose, then radioactivity was subsequently found in the three amino acids. But when the organisms were grown in a medium containing, for instance, unlabelled threonine, then incorporation of radioactivity into this particular amino acid was abolished. The presence of the amino acid thus resulted in a suppression of its continued synthesis – a perfect example of negative feedback. This process, called 'end-product inhibition', means that the biosynthesis of an amino acid is regulated by the amount of the amino acid end-product. As this amount increases, so further biosynthesis is reduced; if its concentration falls, synthesis starts up again. This form of feedback control has since been found to be widespread in biosynthetic systems, in animals as well as micro-organisms, and is probably a major regulatory mechanism.

The situation in this particular example, though, is a little more complicated. The pathway is as follows:

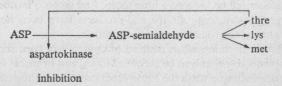

Umbarger was able to show that suppression of synthesis of threonine was caused by the threonine inhibiting the first enzyme along its biosynthetic pathway, i.e. aspartokinase. However, production of the other amino acids was unaffected, even though they too are made from the same precursor, aspartate-semialdehyde. This is because there is not simply one aspartokinase enzyme but three, each with essentially the same enzymic activity but with a slightly different protein structure. Enzymes which exist in multiple form like this are now known to be fairly frequent. They can generally be separated by electrophoresis as they differ slightly in ionic charge, and are known as *isoenzymes*. The existence of multiple isoenzyme forms provides another powerful regulatory mechanism. In the case of aspartokinase, each of the three isoenzymes is specifically inhibited by one of the three amino acids, threonine, lysine, and methionine.

All three must be inhibited before production of the intermediate ceases completely. In addition, each of the amino acids inhibits the first enzyme on its branch line away from aspartate-semialdehyde, so ensuring that the decrease in concentration of the metabolite affects only the production of the inhibiting amino acid. This is an example of control by enzyme multiplicity, although the individual inhibitions are brought about by the type of allosteric processes that have already been described.

A control mechanism that achieves the same effect as this, although by a different mechanism, is the multi-end-product inhibition of allosteric enzymes. Glutamine synthetase (responsible for the production of glutamine from glutamic acid) is regulated by this mechanism. It has binding sites for trytophane, histidine, CTP and ATP among others, because glutamine acts as an amino-group donor in the biosynthesis of all these compounds. If only one of these locks into the synthetase then glutamine production will be turned down by a small amount. If two of them bind, then production will be lowered a little more. And so on. The effect is thus progressive, until all the end-products have been bound, when glutamine production ceases altogether.

An even more ingenious method of control has been studied amongst micro-organisms by Jacques Monod and François Jacob in Paris, which won them the 1965 Nobel Prize. They showed that many micro-organisms, lacking in the enzymes to deal with particular substrates, would, if presented with the substrates, at once begin to synthesize the missing enzymes. In particular, Monod and Jacob studied bacteria lacking in the ability to metabolize the sugar galactose. Presented with galactose as substrate, the bacteria of one strain (though not of others, where even the power of synthesizing the galactose enzymes had been lost) proceeded to fabricate the enzymes needed to utilize it by way of galactose phosphate and UDP-galactose. New enzymes were required, and were synthesized completely from scratch. Similar protein synthesis could be triggered off even if galactose itself was replaced by a non-metabolizable substance which resembled it closely in structure. As the enzymes could be made immediately in response to the substrate, the genetic information necessary for their synthesis must have been present in the cells all along. But the proteins were not normally made. What was the cause of this intriguing effect?

Monod and Jacob argued that this was because in the *absence* of the unusual substrate, the synthesis of the enzymes required to metabolize it was *repressed*, thus ensuring that the cell did not waste time or energy fabricating enzymes for a biosynthetic pathway it would not require.

Induction of synthesis of new enzymes can occur either by stimulating the translation of existing m-RNA or by making new m-RNA molecules by transcription from DNA. Bacterial m-RNA, unlike the m-RNA of higher organisms, has a very short life, half of it decaying within about three minutes, so it seems unlikely that control of translation could have any long-lasting effect. If new m-RNA molecules were made, however, then transcription of the new m-RNA could continue as long as the inducer was present. Geneticists had established that the genes controlling the synthesis of all the enzymes required for galactose metabolism, called the structural genes, were situated along a consecutive length of the DNA, and a little way off was another gene responsible for the manufacture of a specific repressor protein. We have seen that m-RNA is transcribed by a polymerase enzyme that binds first of all to a DNA promoter site. In the absence of the inducer, the repressor protein, according to the Jacob and Monod hypothesis, binds to a site on the DNA between the promoter and the structural genes and thus prevents the m-RNA polymerase from moving along from the promoter to transcribe the DNA. The site to which the repressor protein attaches is called the operator region and the whole system of structural genes, repressor gene, and operator region is called an operon. The inducer exerts its effect by specifically combining with the repressor molecules so that they can no longer bind to the operator and m-RNA polymerase is unobstructed so that transcription can now occur (Figure 26). A cell which can control its protein synthesis so as to make only the enzymes immediately demanded by its metabolic situation – even though it still retained the genetic DNA codes carrying the blueprints for the manufacture of others should this become worthwhile because of an abnormal diet or unusual situation – is clearly highly adaptable and well qualified to survive extreme variations in conditions.

Operon systems have been found to be fairly widespread among micro-organisms. The whole protein synthetic system is, however,

FIGURE 26. *The operon*

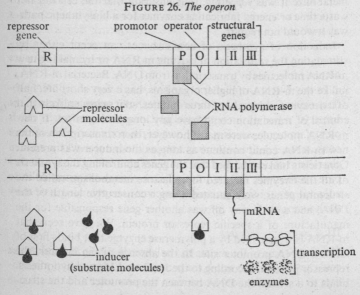

a lot more complicated in multicellular organisms and whether or not operon systems are used here is still a matter for debate. Several different possibilities for control exist in animals because of this complexity. Unlike bacterial DNA, mammalian DNA is associated with the histones and the nucleo-histone complex is much more difficult to transcribe experimentally than is pure DNA. Addition of the acidic chromatin proteins makes transcription a little easier but it seems that in normal chromatin as much as seventy per cent of the DNA is unavailable as a template. This seems to be due to the histones stabilizing the DNA in such a way that makes chain separation by breaking of H bonds more difficult, but the mechanism is highly speculative. Again in animals, unlike bacteria, which do not have a nucleus, m-RNA needs to get from the nucleus where it is made to the ribosomes. The m-RNA in its passage through the cytoplasm may be subject to attack by degradative or modifying enzymes and translation affected in that way, or the rate of passage of the m-RNA may be subject to modification. It is an intriguing though as yet unexplained observation that in the mature mammalian cell, RNA synthesis in the nucleus proceeds at a rapid rate, but very little of the RNA that is made is allowed into the cytoplasm; most is subsequently degraded without ever leaving the nucleus at all. This, too, could provide a sensitive control system helping determine just which m-RNAs are actually translated.

In any event, the need for sensitive regulation of m-RNA production may be much less in the multicellular organism than in the bacteria. After all, in a multicellular organism any given cell is part of an organ which is itself part of the body, has a much more constant and defined environment, and does not meet the sudden changes in condition and diet against which bacteria have to prepare themselves. In the multicellular organisms, such delicate adjustment to changing conditions seems to be provided instead by the hormones.

III. *Control by structure*

These are some of the checks and balances that help to regulate the cell's activity and to relate the production of energy and metabolites to their utilization. There must be many others, for this is one of the newer aspects of biochemistry, in which both experimental techniques and theoretical tools are lacking. At least one other major factor in cellular control can be predicted, though, and this is the control implicit in the very structure of the cell itself. We have described how the cell is built up of a series of subcellular organelles, nucleus, mitochondria, ribosomes, surrounded by cytoplasm and bounded by the cell membrane. Each organelle contains characteristic enzymes and substrates, and those present in one organelle are not necessarily freely available to those in another – the dangerous hydrolytic enzymes, for example, are kept firmly caged within the lysosomes, and allowed out only under extreme conditions. Thus the cell can be seen as consisting of a series of separate 'compartments' bounded by walls through which some substances but not all can pass. Across these walls considerable differences of condition may exist; differences of acidity and alkalinity, temperature, richness or poorness in ATP or in essential cofactors such as magnesium. It may not be assumed that, because the overall concentration of a substance within the cell is high, it is present in the same high concentration in all parts. Some regions may have none at all. And even apparently slight differences may have formidable consequences. A temperature difference across a mitochondrial membrane of only 0·01 degrees Centigrade does not sound much, but so thin is the membrane that it represents a temperature *gradient* of ten thousand degrees per centimetre. Under such conditions many hitherto unsuspected control systems may function.

What *is* known is that the cellular membranes are extremely fussy about what they will and will not allow to pass through them; as has been discussed in Chapter 4, even quite small molecules may have their free diffusion into and out of the cell and its compartments restrained by the membrane. Mechanisms of facilitated diffusion and active transport, which are controlled by the membrane, are involved in these processes – helping, for instance, to keep the interior of the cell in a high K^+, low Na^+ condition, despite the fact that the extracellular environment – for example, the blood and plasma in mammals – is maintained at high Na^+, low K^+ levels.

Similarly, membranes within the cell show selectivity between various cell components. The mitochondrial membrane appears to have almost a vacuum-cleaner avidity for the cell's calcium, which is sucked up and stored within the mitochondrion to an amazing extent. Such phenomena must be concerned with control mechanisms in ways we do not even begin to understand as yet. We can, however, see how they can be used in principle to provide a fine control of cell metabolism. One last example must suffice, drawn from the energy-regulating systems. We have shown that nearly all the cell's ATP is formed within the mitochondria. Now there is good evidence that the mitochondrial membrane may present a barrier to the free flow of ATP to other parts of the cell, and of ADP from outside into the mitochondrion, for rephosphorylation. Many reaction pathways have as a rate-limiting factor the availability of ATP – one of the best examples is the glycolytic breakdown of glucose itself to pyruvic acid. This reaction pathway has as its initial, and critical, step the phosphorylation of glucose by hexokinase, and almost immediately following, a second phosphorylation, of fructose-6-phosphate to fructose 1, 6, diphosphate (reactions 1 to 3 of Chapter 8). Study of the kinetics of these two reactions has shown that they are the slowest, and therefore the rate-limiting ones, of the whole glycolytic sequence. In the presence of glucose, the limiting factor is the availability of ATP which is itself largely the product of the oxidation of the metabolites of glucose. Glycolysis ought then to provide an example of positive feedback, as shown in Figure 27(a).

But in fact, one of the oldest observations in biochemistry, named for its discoverer as the Pasteur effect, and first found in yeast but later extended to other tissues, was that the addition of glucose to a respiring tissue *diminishes* respiration, whilst increasing the respiration of a glycolysing tissue *diminishes* glucose utilization. A century after Pasteur, it would seem that the solution to this apparent paradox may perhaps be found in the fact that, in some way, mitochondrially produced ATP from pyruvic acid oxidation is *not available* for the glucose in the cytoplasm, whilst ADP produced during the hexokinase reaction is *not available* to the mitochondrial phosphorylating sites. Rather than the positive feedback of Figure 27(a) the more complex pattern of negative feedback of Figure 27(b) seems to apply.

FIGURE 27.

The Pasteur effect: an example of control by structure

(a)

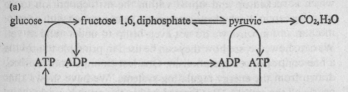

(a) If there were no other obstacles involved, glucose oxidation would be an example of *positive* feedback.

(b)

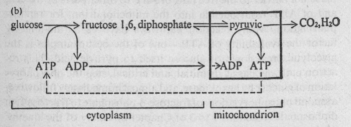

(b) But when the membranes are put in, the picture becomes more complex; free movement of ADP and ATP is blocked and the result is *negative* feedback.

But these studies are in their infancy. The full implications of the role of structural barriers within the cell will most probably await a future generation of biochemists to unravel; for the moment, we may merely speculate.

EXTERNAL REGULATION OF CELL METABOLISM: THE HORMONES

As well as operating as an internally self-regulating, or *homeostatic* mechanism, the animal cell forms part of a larger whole. It is one of many hundreds of millions in a particular organ, the organ one of a group of interdependent parts of the body. The survival of the cell is not the ultimate goal; rather the function of the cell is subordinated to the survival of the animal, and, by reproduction, of the species. The animal, then, must have means of interlocking and co-ordinating the functions of the various cells of the body so that they can unite to respond to varying conditions and stimuli in a uniform, constructive manner. They may be called upon to secrete saliva and enzymes for digestion, to contract in order to move a limb, to synthesize added amounts of glycogen, or to release more glucose into the blood stream. Many of these functions are under the direct control of nerves, which carry specific messages from the brain or the spinal cord to the individual organs of the different parts of the body. The role of these nerves is the province of the physiologist, although we shall be able in the next chapter to consider a biochemical explanation of *how* they carry messages. But a large number of cellular activities are also regulated by means of a series of special chemicals, synthesized in a group of small organs scattered through the body, and released from these organs, in response to certain stimuli, to be carried through the bloodstream to all parts of the body. These chemicals carry to their 'target cells' special messages of instruction, intended to stimulate, or suppress, or otherwise modify their activities. This group of messenger-chemicals is called 'hormones' (from the Greek for 'I arouse to activity'); the organs which initially secrete them are known as 'endocrine glands'.

The endocrine glands are permeated by a rich network of blood capillaries which allow the hormones to be secreted directly into the circulatory system. All the hormones have certain things in common, and in terms of chemical composition fall into two main groups, the peptide hormones and the steroid hormones. Their synthesis within the cells of their endocrine gland usually involves the prior formation of a pro-hormone molecule which is physiologically inactive and is only converted to the active form when it is needed. They are nearly all stored within the cells that synthesize them, packaged in membrane-bound granules formed by the Golgi complexes that exist in large numbers in these secretory cells. Secretion is often mediated by other hormones or by nervous control and the circulating levels of several of the hormones fluctuate over particular time periods (for instance, daily, monthly) with precise regularity. They are nearly all transported in the blood complexed with plasma proteins, whose function is to protect the hormone from degradative attack until they are safely at their targets, and they also serve as a reservoir for the hormone as these protein/hormone complexes are in equilibrium with free hormone. As the level of free hormone drops so more hormone is dissociated from its protein carrier to maintain the equilibrium.

The hormones enable the body to exert subtle control over a host of metabolic functions and to alter these functions in response to changes in conditions in the outside world. In general, the difference between the jobs of nerves and hormones may be summed up like this. The nerve commands a specific cell or group of cells to take a particular, specific, limited action – a muscle to contract, the eye to change focus, or a secretory cell to release its secretion. The hormones, on the other hand, play the role of *slightly* modifying the normal behaviour of the cells of their target organ so as to cause a relatively small but often prolonged alteration in its pattern of activity. It is rather like the difference between the wavelength and the tone control on a radio – the nerve chooses the particular wavelength on which the cell should function, the hormone alters its tone to keep it in tune with the best needs of the rest of the body.

Such a distinction also means that there is a dramatic difference between a person whose nerves cease to function and one whose hormonal control gets out of balance. Blocking the nerve to a limb or organ leaves it paralysed and useless, whilst removing one of the hormone-secreting glands tends to leave the body capable of functioning after a fashion, but with a reduced ability to rouse itself to meet altered circumstances. Minus one of its hormones, an animal will perhaps put on an excessive amount of weight, or be unduly sensitive to cold, or be subject to sexual irregularities. Such differences are difficult to define, but very obvious on examination. It may be that hormones represent, evolutionarily speaking, an earlier and more primitive attempt at co-ordination and control of the body's activities than that represented by the nerves.

At any rate, circulating in the bloodstream of the human there may be as many as fifty different chemicals, which, between them, are capable of causing substantial differences to the metabolic pattern of the target cells which respond to them. We cannot here describe all the hormones or all their functions; in many cases they are extremely complex, and only in the last few years have some of the mechanisms begun to be elucidated. In general it seems a sensible assumption that all hormonal activity can ultimately be related to the effect of the hormone in altering the velocity of some particular rate-limiting enzymic reaction, but the hormones are never found free within the cytoplasm and their interaction with the target cell is by way of highly specific receptors.

There are a number of ways of grouping hormones. The simplest is according to the glands that secrete them. Thus we have *thyroxine*, the hormone of the thyroid gland of the neck, and *parathormone*, made in the tiny parathyroids which lie at each side of the thyroids. *Insulin* and *glucagon* are secreted by the pancreas. Above each kidney there is a small gland called the adrenal, the central region of which produces *adrenalin* whilst the outer regions (the cortex) produce a battery of *steroid hormones*: *aldosterone, corticosterone*, etc. The testis and ovary produce their respective sex hormones, *testosterone* and *oestrogen*. Finally, a minuscule organ, in humans the size of a pea, located in a cavity of bone between the bottom of the brain and the palate at the top of the mouth, produces up to twenty-five or more different hormones, which have the supremely important job of controlling the output of virtually all the other hormones in the body. This master-controller is the *pituitary*.

But here we are interested only in the biochemical role of the hormones as cell regulators, and will look at the general features of the control processes with which they are concerned. The results of a slight change in the balance of cell metabolism may differ profoundly for the animal depending on whether the cell concerned is, say, in the liver, the brain, or the testis. Although the physiologist may find in one case the onset of fatty liver, in another coma and mental aberrations, in a third abnormal sexual development, the biochemical mechanics behind all three may be virtually identical. Physiologically, the sex hormones are responsible for the control of puberty and menopause, ovulation, menstruation, pregnancy, parturition and milk production, and secondary sexual characteristics such as breasts, to say nothing of involvement 'psychological' level of sexual drives themselves. Yet to the biochemist none of these more interesting features is relevant (at least during working hours) and here we shall concentrate on the general regulating features of peptide and steroid hormones as research over the last decade has revealed them.

Peptide hormones and second messengers

Peptide hormones are known to interact with membrane-bound receptor molecules in their target cells. It is only with the development of new techniques for the isolation and characterization of these receptors and methods for investigating hormone-receptor interaction that it has become possible to hypothesize about the mechanism of hormone action. Essentially the binding of hormones to receptors turns out to be analogous to the binding of enzymes to substrates or prosthetic groups. Many of the techniques for receptor study are already familiar from earlier chapters. They include labelling the receptor with a radioactive molecule similar in structure to the hormone that will bind to it. Such a substance is termed a ligand and may even evoke the same physiological response as the hormone, in which case it is called an agonist. Antagonists are substances that bind to the receptor but do not evoke a physiological response and are widely used in medicine as inhibitors of hormone action. Radioactive labelling makes it possible to determine the kinetics of the ligand binding reaction and to show that binding is usually very rapid but reversible and that bound hormone establishes an equilibrium in time with free hormone. Because receptors are present in low concentrations, their isolation from a tissue homogenate is extremely difficult although made possible by the use of various density gradient centrifugation and chromatographic methods. Nearly all the receptors that have been characterized have turned out to be proteins or glycoproteins, and this helps explain perhaps the most important and dramatic property of receptors – their ability to discriminate between very similar ligands and to specifically bind the one hormone that structurally interacts with the binding site. The mechanism by which the receptor-hormone interaction is translated into a response within the cell is the subject of current intensive research and although new and exciting information is steadily emerging it is only possible at the moment to give a brief outline of the stages involved. These can best be summarized with the diagram of Figure 28:

FIGURE 28. *Mechanism of hormone action*

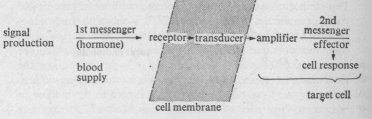

We have already discussed the receptors and we have said that there are only a limited number of each type on the cell surface. What is more, kinetic studies show that any one receptor molecule is unlikely to bind more than one hormone molecule. But these small numbers of interactions somehow trigger an enormous and often long-lasting response within the cell. This led to the conclusion that the hormone signal must be amplified in some way by the cell. One way of performing this amplification would be for the hormone (first messenger) to trigger the production of another, and it is this second substance (second messenger) that then affects the intracellular target substance, known as the effector. In the late 1950s Earl Sutherland and his co-workers in the United States isolated a compound that was capable of mimicking the effects of many hormones when administered alone to samples of target tissues. This compound was identified as a form of the nucleotide AMP in which the molecule is bent back on itself as a ring, cyclic AMP, which has been found to fill the role of second messenger in a vast number of cell-hormone interactions.

Cyclic AMP is formed from ATP by the enzyme adenylate cyclase, which is situated on the inner cell membrane, and it is this enzyme that is responsible for the amplification of the hormone signal, because just one enzyme molecule, when activated by the hormone-receptor complex, can of course produce many cyclic

Formula of cyclic AMP

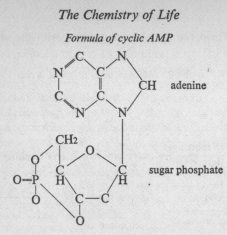

AMP molecules. Adenylate cyclase is an allosteric enzyme with a strict requirement for Mg^{2+}, whose binding enhances the subsequent binding of ATP during the reaction

$$ATP \xrightarrow[\substack{\text{adenylate} \\ \text{cyclase} \\ Mg^{2+}}]{} cAMP + \text{pyrophosphate}$$

Even though the adenylate cyclase is membrane-bound, the experimental evidence suggests that it may be spatially separated from the receptor; and therefore some kind of transducer mechanism is brought into play when the hormone-receptor complex is formed. Several explanations of transducer role have been advanced. In isolated tissue systems, guanosine triphosphate (GTP) and other guanyl nucleotides stimulate adenylate cyclase, so it has been suggested that hormone binding could be linked to GTP production or activation within the membrane. Another model is based on the observation that many hormones are composed of two subunits. One of these subunits might then be responsible for receptor recognition and is therefore hormone specific. The other subunit would be the same in all the hormones and it could be that this subunit is released on binding and it in turn binds to the adenylate cyclase, allosterically enhancing its activity (see Figure 29). A third explanation is based on the experimental finding that hormone binding increases membrane fluidity by affecting the structure of the lipid phosphatidylinositol. Increased fluidity enables the hormone-receptor complex to move within the membrane, until it encounters an equally mobile adenylate cyclase molecule and causes its activation. According to this model, free receptors without the bound hormone will be in the wrong conformation to interact with the enzyme.

FIGURE 29. *Possible model of transducer mechanism*

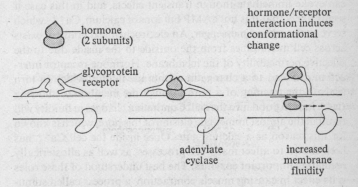

Cyclic AMP has many diverse effects on cell metabolism. In most cases its action is dependent on its capacity to activate a class of enzymes known as protein kinases. Protein kinases, when activated, are able to phosphorylate other proteins (generally by adding a phosphate group to one or more of the serine residues present along the amino acid chain of the effector enzyme). The phosphorylation then alters the effector enzyme's receptiveness towards its substrate. We shall see an example of how this works when we go on to discuss the individual hormones.

The method of protein kinase activation by cAMP is fairly well established but, more recently, information has emerged suggesting that at least in bacterial cells, the nucleotide also has a direct role in regulating the transcription process. It is thought to do this by binding to the DNA at the promoter site, complexed with a carrier protein. This binding induces a conformational change in the DNA, making it easier for m-RNA polymerase to bind and initiate transcription. The effect of cAMP at the nucleus in the cells of higher organisms is much more obscure but it does seem to have some effect on the histones and acidic chromatin proteins which, as we saw in the last chapter, could influence transcription. When cAMP is involved as a second messenger, the effect of the hormone tends to be a large and long-lasting response, but because of the several steps involved a delay often occurs between receptor binding and the ultimate cellular response. Some hormones, however, can evoke immediate though transient effects, and in this case it seems likely that it is not cAMP but ions of calcium, Ca^{++}, which serve as the second messenger. An electrochemical gradient exists across cell membranes from the outside to the inside due to the selective permeability of the membrane. Hormone-receptor interaction can lead to a change in membrane fluidity which in turn leads to the opening of a pore through the membrane, allowing Ca^{++} ions to flood into the cell. Continuing changes in fluidity will result in the almost immediate closing of this pore, which is known for this reason as a calcium gate. Once inside the cell Ca^{++} has been shown to affect many cell processes, as well as allosterically regulating important enzymes. The best understood of these roles is its effect in causing muscle contraction, a process called stimulus-contraction-coupling (see the next chapter). Ca^{++} has also been implicated in the secretion of packaged cell products such as hormones and digestive enzymes. It is also thought to be involved in the process of cell division, an effect known as stimulus-division-coupling. The whole picture of hormone action is complicated by the fact that many cells have receptors on their surface for different hormones, linked to either of the two second messenger systems. In cells where both types of systems do exist they have been shown to act antagonistically towards each other, providing a complex control mechanism for modifying the cellular response to hormones. For instance, Ca^{++} has been shown to allosterically activate the enzyme phosphodiesterase which is responsible for the hydrolysis of cyclic AMP. Conversely cyclic AMP is thought to speed up the membrane pumping mechanism that is responsible for reducing the intracellular Ca^{++} levels.

Steroid hormones

The steroid hormones can be subdivided into three classes depending on the processes they control:

(1) Glucocorticoids which help regulate glucose metabolism.

(2) Mineralocorticoids which help maintain the ionic balance within the body.

(3) Sex hormones.

Their mode of action differs fundamentally from that of the peptide hormones in that as steroids they are hydrophobic molecules, they can penetrate the membrane and therefore interact with intracellular receptors in the cytoplasm rather than membrane-bound receptors. So the steroid hormones do not need to use a second messenger to evoke a response. When the cytoplasmic receptor is saturated with hormone the whole complex moves into the nucleus and binds to the chromatin, causing a direct stimulation of m-RNA production by facilitating the action of m-RNA polymerase. Binding is thought to be due not to DNA itself but to the acidic chromatin proteins, although with at least one of the sex hormones (progesterone) this leads to a dissociation of the receptor-hormone complex into two subunits, one of which remains bound to the acidic nuclear protein and frees the other to subsequently bind to DNA. Thus it is the interaction between the nuclear protein and the first subunit that is responsible for specificity, for recognizing the correct place on the DNA strand at which transcription should be initiated. The basic histone proteins may have a controlling influence here, by masking the binding sites on the acidic proteins under certain conditions.

HORMONE ACTION

Energy metabolism

Thyroxine. The thyroid gland of the neck secretes a hormone whose primary function seems to lie in a regulation of the body's energy metabolism. Those whose thyroid gland is deficient suffer from *hypothyroidism.* They are slow, soporific, and apparently lazy; they put on weight but never seem to have the energy to act. If the disease occurs in early childhood, the infant becomes a *cretin,* dwarflike and mentally retarded. People with an overactive thyroid (*hyperthyroidism*) are by contrast thin and over-easily excitable; their eyes 'pop' and they are generally 'twitchy'. Both these groups of symptoms are related to a function which is termed 'basic metabolic rate' (BMR) which is the rate at which an individual burns up food to provide energy under resting conditions. Thyroid deficiency lowers BMR from a standard 100 to 50 or below; hyperthyroidism raises BMR to 150 or more.

The hormone made by the thyroid which is responsible for regulating BMR is called *thyroxine*; it is an iodine-containing amino acid derived from tyrosine. Almost the whole of the body's stock of iodine is used by the thyroid in the manufacture of thyroxine, which has the structure

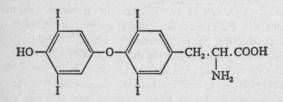

and another reason for thyroid deficiency diseases is a diet low in iodine. When an individual has to exist for a long time on such a diet – as, for example, used to happen in various areas of Britain where drinking water was abnormally low in iodine – the thyroid swells to make use of every last scrap of available iodine, a condition known clinically as goitre, and once colloquially described as 'Derbyshire neck'.

Thyroxine is one of the few hormones for which unequivocal effects on isolated biochemical systems can be observed; in particular, it has been shown that thyroxine at high concentrations has a direct effect on mitochondrial oxidation. Tissues from animals which have been injected with thyroid extracts or purified thyroxine show a raised oxygen consumption, particularly in liver and muscle. Similarly the addition of thyroxine to tissue slices of the same organs results in increased oxidation. Thyroxine seems to be especially trapped by the mitochondria, where it stimulates the activity of succinic dehydrogenase. One, but by no means the only, explanation for the role of thyroxine is that it acts in some way as an 'uncoupler' of oxidation from phosphorylation. Uncoupling, it will be remembered, is a method of slipping the oxidative phosphorylation system into neutral by disconnecting it. When this happens, oxidation can take place in the absence of phosphorylation. It is probable that under normal conditions the mitochondrial oxidation system is partially uncoupled in this way so that excess foodstuff can be burned away with the dissipation of its energy as heat but without being used to make ATP. In the absence of thyroxine perhaps the degree of coupling becomes higher, so that no food at all can be oxidized without ATP production. The foodstuff in excess of needs will then all be stored, the individual's BMR will slow down, and he will grow obese. Thyroxine excess, leading to considerable uncoupling, would mean that nearly all the food taken up would be burned, little would be stored, and the individual's BMR would go up and his weight down.

Just how thyroxine acts to regulate uncoupling is not known. Probably it does so by some interaction with the mitochondrial membrane, rather than directly with an enzyme system. Electron microscope pictures of mitochondria treated with thyroxine show characteristic changes; they are very fragile and begin to swell up like tiny balloons. These observations are presumably connected in some way with the activity of the hormone, for such changes in the membrane must affect the enzyme systems which are built into it. It should be emphasized, though, that with thyroxine, as with virtually all other hormones, there is more than one school of thought as to mechanism. For example, in contrast to the picture of thyroxine as an uncoupler which we have just presented, there are those who maintain, with good experimental evidence to support them, that one of the effects of thyroxine is to speed up protein synthesis, and, in particular, the synthesis of mitochondrial enzymes which are presumably concerned with oxidation and phosphorylation. By increasing the amounts of such enzymes, thyroxine would presumably speed up the rate of oxidation in the mitochondria, providing that ADP and (P) concentrations were not rate-limiting.

The Chemistry of Life

The regulation of carbohydrate metabolism

A variety of hormones affect the body's carbohydrate metabolism; some increase carbohydrate reserves by promoting the synthesis of glycogen, others accelerate the rate of glycogen breakdown and glucose oxidation. Between them, they add formidable weapons to the cell's already powerful armoury of glucose control systems.

The hormones mainly responsible are produced by two endocrine glands, the pancreas and the adrenals. The pancreas is a large gland situated in the duodenal loop of the intestine and ninety-five per cent of its tissue is devoted to the secretion of digestive juices which are carried to the intestine by the pancreatic duct. The other five per cent of the tissue (the so-called islets of Langerhans) contain several distinct cell types. The α-cells are responsible for the production of the hormone glucagon whilst the β-cells make insulin. Insulin is a low-molecular-weight protein which we have already met as the first ever to have its full amino acid sequence determined. In patients suffering from insulin deficiency (usually through failure of the insulin-producing β-cells), glucose circulates in large quantities through the bloodstream but is not taken up by the muscle or liver either for oxidation or for conversion into glycogen. Ultimately, it is disposed of by excretion into the urine. Thus the patient suffers from a diminished carbohydrate metabolism despite the existence of a plentiful supply of glucose, and the body has to resort instead to fats and proteins to make up the energy-deficiency (resulting in the condition known as 'fatty liver' – see page 161). The insulin-deficiency illness is known as diabetes.

It was Banting and Best in Canada who in 1921 were able to prepare a pancreatic extract containing insulin and to demonstrate that its injection into diabetic patients resulted in an immediate uptake of glucose into the tissues and a normalization of carbohydrate metabolism. Excessive insulin, on the other hand, results is so great a flow of glucose into liver and muscle that other organs, and in particular the brain, actually become glucose starved. Such starvation rapidly produces coma and even death, if not immediately alleviated by massive glucose injections. Insulin production must therefore be carefully balanced to avoid either too little or too great an uptake. Physiologically, insulin thus offers one of the clearest defined of all hormonal effects, and one whose mechanism of action is now beginning to become clearer.

Adrenalin is produced by the central core region (medulla) of the adrenal gland. The hormone is secreted as part of the body's response to a new or potentially dangerous situation – adrenalin has been called the 'fight or flight' hormone.* Under its influence the heart beats faster, more blood flows to the muscles, several of which begin to contract, and the condition which we recognize in ourselves when we prepare to meet an emergency – a contracted stomach, muscular tension, and general 'nerves' – is produced.

A variety of these effects of adrenalin are mediated through the nervous system in ways that take us outside the scope of this book to discuss. So far as glucose metabolism is concerned, its role is to stimulate glycogen *breakdown* in muscle and liver. In this respect its function is similar to glucagon although the two differ in structure, as glucagon, like insulin, is a peptide. Insulin, as we have already said, has an opposing role to the other two hormones, in stimulating glycogen *synthesis*. These hormones work in a concerted fashion via the cyclic AMP and protein kinase system. They modulate this system by modifying either the glycogen phosphorylase or glycogen synthetase enzymes. Both the phosphorylase and the synthetase exist in two forms, an active and an inactive state. In each case the hormone exerts its effect by phosphorylating the enzyme, the difference being that phosphorylation results in the production of the *active* form of glycogen phosphorylase but the *inactive* form of glycogen synthetase.

There are thus three consecutive steps involved in the process leading to, for instance, the stimulation of glycogen synthetase following hormone-receptor interaction. Each step magnifies or amplifies the effect of the preceding one and the whole reaction is therefore known as a cascade system. It is illustrated in Figure 30.

If this seems complex, it is still only a simplification of what appears to happen *in vivo*. The enzyme responsible for phosphorylating the glycogen phosphorylase, phosphorylase kinase, is *itself* capable of existing in an active and an inactive form, and there

* Actually 'fight, flight or frolic' but the puritanical English tradition usually omits the last.

should now be no prizes for guessing that the difference once again lies in the state of phosphorylation of the phosphorylase kinase. So there is yet another enzyme which activates phosphorylase kinase by phosphorylating it, and one which deactivates it by dephosphorylating it!

The interplay of these enzymes and their hormonal regulators by mutual phosphorylation and dephosphorylation constitutes one of the most complex and best studied control mechanisms in mammalian systems. The whole process is summarized in Figure 30, which also shows how insulin acts by inhibiting both pathways simultaneously.

FIGURE 30. *Hormonal control of glycogen metabolism – cascade system*

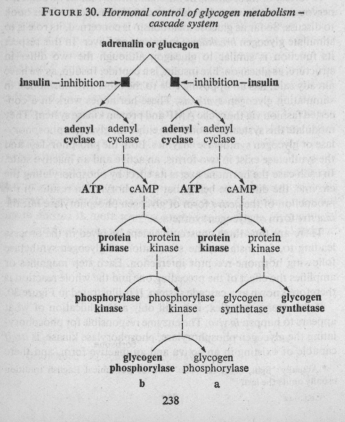

238

It is worth emphasizing that one of the important differences between the two hormones, and also between adrenalin and thyroxine, which also stimulates metabolism, is in part the immediacy of the action of adrenalin; the adrenal gland releases it in brief bursts which cause sudden swift changes in the balance of glucose metabolism, but these effects rapidly diminish as adrenalin itself is soon destroyed by specific oxidizing enzymes, especially in the muscle. Thyroxine, insulin, and glucagon have long-term regulating roles, whilst adrenalin is for the here and now of sudden need. Incidentally, whilst the cascade regulating systems invoking phosphorylation and dephosphorylation of enzymes, interaction with cAMP and so forth have proved a very fertile field of study in recent years, it is not necessarily the case that all the actions of the hormones can be accounted for in terms of this interplay. The hormone literature is full of other earlier observed effects of the hormones on various biochemical systems, not all of which seem to be fully describable as secondary to the glucose regulating steps. What they do show is the extreme sensitivity of the whole metabolic process to the entry of glucose into its metabolic pathway, and hence the subtlety of control systems which have developed over evolutionary periods, to regulate it.

Glucocorticoids. The adrenal gland, or rather its outer margin, the cortex, also secretes another group of hormones that play a significant role in the regulation of carbohydrate metabolism, the glucocorticoids, a name given to a group of related substances (including *cortisone* and *cortisol*) which chemically is characterized by having an alcoholic or ketonic oxygen atom at carbon atom number 11 of the steroid molecule, shown for cortisone here:

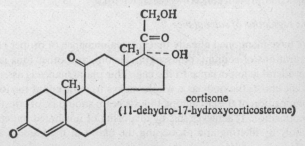

cortisone
(11-dehydro-17-hydroxycorticosterone)

The main effect of these hormones is to increase the laying down of glycogen in the liver. At the same time, when injected into an experimental animal they put it into negative nitrogen balance – that is, they increase the breakdown of proteins and amino acids. In general, they seem to operate at least partly so as to reverse or counterbalance the effects of insulin, so that, in a normal healthy animal or human, carbohydrate metabolism is controlled partly as the result of a tug-of-war between insulin on the one hand and the glucocorticoids on the other.

The fact that the glucocorticoids act so as to increase both glycogen synthesis and protein breakdown would suggest that they worked by speeding the conversion of amino acids to carbohydrate, and in fact this has actually been demonstrated by studying the fate of radioactive amino acids under the influence of the hormones. Glucocorticoids were found to have a direct effect in increasing several-fold the amounts of certain enzymes of amino acid breakdown, and in particular of *transaminase* and *amino acid oxidase* of the liver. The result of increasing the activity of these enzymes would be to increase the rates at which amino acids were converted into urea and tricarboxylic-acid cycle intermediates such as α-oxoglutaric acid. The urea is excreted and the α-oxoglutaric acid can be *either* oxidized *or* used to resynthesize glucose and glycogen by pathways we have already described in detail.

Thus these hormones are particularly important during prolonged starvation when glucose reserves are used up and the cell has to turn at last to the proteins. The amounts of transaminase and related enzymes are of course increased in the way already described, by a specific initiation of m-RNA coding for the particular protein concerned (Chapter 10).

The regulation of salt balance

We have mentioned already that the maintenance of proper concentrations of sodium, potassium, calcium, and other ions is of considerable importance to the cell. This maintenance is assisted by the efforts the body as a whole makes to ensure that the ionic composition of the circulating bloodstream should be of constant composition. It achieves this by disposing of unwanted material, mainly by filtering and processing the blood in the kidneys and excreting excess and undesirable substances through the urine. One of the principal substances disposed of in this way is nitrogen-ammonia, which is got rid of as urea, but the urine also contains varying amounts of many other inorganic substances as well as some organic ones, in addition, of course, to water. Several hormones regulate the balance struck by the kidneys between quantities retained and disposed of. Vasopressin is an antidiuretic hormone released by one region of the pituitary and controls the amount of water released as urine. The mineralocorticoids, the best known of which is aldosterone, help diminish the loss of sodium, chloride, and bicarbonate ions (Na^+, Cl^-, HCO_3^-) and increase the excretion of potassium and phosphate (K^+, HPO_4^-). Calcitonin, produced by the thyroid, inhibits calcium release from bones and therefore decreases blood Ca^{++} levels. The parathyroid hormone, *parathormone*, has an opposing effect, as it stimulates the excretion of calcium and phosphate (Ca^{++}, HPO_4^{--}) and achieves this at the expense of the major deposits of calcium phosphate in the body – the bones. All these hormones probably exert their effects by altering the properties of the filtering systems in the kidney cells which are responsible for selecting which substances to retain in the blood and which to release. The specific biochemical mechanisms concerned, though, are very far from being understood.

Regulation of protein metabolism

Indirectly, many hormones influence protein metabolism. Thus hormones like the glucocorticoids which encourage glycogen synthesis, do so at the expense of proteins and amino acids and put the body into negative nitrogen balance. Similarly, insulin deficiency, by demanding that energy be provided at the expense of substances other than carbohydrate, encourages protein catabolism whilst addition of insulin has the reverse effect. But these happenings are probably subsidiary to other, primary events. The hormone which seems to be most directly concerned with regulating protein synthesis is itself a pituitary protein called *growth hormone*. Its presence seems essential for normal growth in the infant and adolescent, and, in its absence, *pituitary dwarfism* may result; these dwarves, it may be noted, unlike cretins, are not at all mentally retarded, but simply undersized, reaching no more than three or four feet in height. Similarly, excessive growth hormone production in youth produces pituitary *gigantism* – giants who may be seven or even eight feet tall. Although growth hormone is produced right through life, its effects are most marked during the rapid growth of infancy; after puberty its activity seems to become modified, probably by other hormones which counter its activities. Some idea of the scale of its effects in youth may be gained by considering an experiment where, in rats whose pituitaries had been removed, the daily injection of as little as 0·01 milligrams for nine days resulted in an increase of 10 grams in body weight. Growth hormone works by stimulating the liver to produce particular proteins called somatomedins which induce specific m-RNA synthesis, although there is also some evidence to show that it also directly increases the rate of protein synthesis of the ribosome.

Regulation of other hormones

We have described a variety of regulatory influences exerted by a number of different hormones, which between them are capable of modifying a substantial range of cellular activities. Several of these hormones, indeed, have mutually contradictory effects, and the final state of the cell must depend on the interplay between them and the relative amounts of each present. Either overproduction or underproduction of a hormone may lead to undesirable consequences: deformity, disease, coma, or death. Clearly, then, there must be some means whereby the healthy body can regulate the regulators, ensuring that the quantities of each hormone present at any one time are appropriate to the metabolic demands of the whole animal.

This regulation is achieved by a device of breath-taking simplicity. As each hormone is itself the product of biosynthetic pathways in particular endocrine organs, the amount present in the bloodstream may be controlled by an alteration of the rate at which it is being synthesized. And this rate is controlled for nearly all the major hormones (the big exception is insulin) by *other* hormones, produced in the pituitary. Thus the pituitary makes a set of peptide hormones whose target organs are the other endocrine glands of the body. These secondary higher level regulators are called the *trophic* hormones (from the Greek verb 'to nourish'); they include thyrotrophic, adrenocorticotrophic, and gonadotrophic hormones, amongst others, and each has the property of stimulating the synthesis and release of the primary hormones from their respective glands.

But there is a possible dangerous trap here. What regulates the production of the pituitary hormones? Are we in danger of demanding an infinite regress of hormones, as once we found ourselves involved with an infinite regress of enzymes? There are two parts to the answer to this question, which between them reveal the full beauty of the regulatory mechanisms which the body has evolved. One is that the pituitary gland itself lies at the base of the brain and is thus at least partially under nervous control. The second is that experiments show that the rate of production of the trophic hormones themselves is dependent on the concentration of the hormones of their target glands.

For example, during the female reproductive cycle, the development and mobilization of the egg cell (oocyte) is controlled by two hormones produced by a particular region of the pituitary, Follicle Stimulating Hormone (FSH) and Luteinizing Hormone (LH). LH also stimulates the follicle cells around the developing ooctye to produce oestrogens which help prepare the placenta for pregnancy and which also inhibits FSH. If conception occurs, then the placenta takes over oestrogen production, ensuring continued inhibition of FSH and therefore making sure further ovulation does not occur. If there is no pregnancy the developing placenta regresses and is shed during menstruation. The follicle cells regress at the same time so that oestrogen levels decrease, and FSH levels rise, initiating another ovulating cycle. FSH is not only subject to inhibition by oestrogen but also comes under the control of an FSH releasing hormone (FSH-RH) produced in another part of the pituitary. FSH-RH is also subject to feedback inhibition by oestrogen and FSH is in fact controlled by the combined effect of the two hormones.

Similarly the presence of thyroxine inhibits thyrotrophic hormone production and thyrotrophic hormone stimulates thyroxine production. This can be shown diagrammatically as a feedback loop (Figure 31) albeit under ultimate higher control from the brain. Hormone production thus forms a self-regulating system efficient and simple enough to gladden the cyberneticist's heart.

At the start of the chapter we showed how we could describe the cell as an automaton by using the knowledge of the simple laws of chemistry and physics and the logical tools provided by the cyberneticists. But we recognized that we could not go on regarding the cell in isolation. Cells are organized into tissue systems and organs, each performing a particular task for the general well-being of the body. Not only do the cells in any one organ have to function together, they must also be aware of and responsive to events happening in other parts of the body. This led us to a discussion of the hormones by which the body controls and integrates the metabolism of its individual cells. So we came to a view of cells as interacting units in a greater whole, the entire organism.

FIGURE 31.

Hormonal feedback – control of hormone production

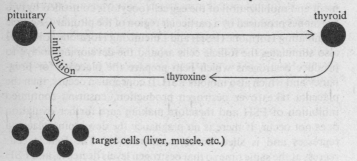

In studying the finer aspects of the regulation of the cell, we found it necessary to take into consideration what is, for the cell, the outside world of the body as a whole. Now that we have come so far as to begin to build up a biochemical picture of the whole body, it becomes necessary for us to take one final step. For the cell, the external environment is provided by the body, a homeostatic system which maintains the constancy of its internal composition and thus enables the cell to survive as part of an integrated whole.

For the cell, departure by very much from the norms of this uniform environment spells speedy death.

But the body also has an external environment, and this, far from being the warm dark womb in which it cradles its own cells, is a tough, highly changeable, and often dangerous jungle. It is a jungle of rapid alterations in temperature, humidity, and consistency, in which food, instead of being wafted effortlessly along a swift-flowing bloodstream, must be hunted down ruthlessly and captured by skill. The body needs to maintain homeostasis, but in order to do so it has to convert the non-homeostatic systems of the world around it. It has to *act* on this world in order to mould it into favourable contours. If temperature changes, the organism must organize itself so as to reverse these changes, by warming its surroundings or moving to more sheltered ones. It must be constantly in search of food, of air, of water, in order to meet the insatiable demands of its own interior.

And in this search it must inevitably conflict with other life forms which are driven by the same internal urges to hunt for the same ends (for instance, organisms which are bent on devouring *it*). This need to act upon the world around it demands a biochemical and physiological integration of the body at the highest level, and it is to that which we must now turn if we are to achieve the biochemist's goal of describing the *total* behaviour of living organisms in physical and chemical terms.

CHAPTER 12

THE CELL IN ACTION

For the organism to act on the world around it, three abilities are required. In the first place, it must have the means of receiving stimuli from outside which will enable it to assess its position in space and its relationship to other objects; it must be able to see, hear, scent, and feel, food and danger; i.e. it must have *sense receptors*, eyes, ears, nose. Second, it must be able to recognize, coordinate, and respond in the correct way to incoming sense stimuli. This is the job of the brain and central nervous system. Third, it must have a means of achieving purposeful movement in order to obtain a given goal, whether the capture of food or the avoidance of harm; this movement is produced by *muscles*. We can sum up this three-cornered relationship in a simple diagram (Figure 32).

A considerable percentage of the human body is taken up by these specialized organs – sensory receptors, limbs, and brain – and most human life at the conscious level is lived only in their terms; we are content to leave our internal homeostasis to operate by means of nerves, hormones, and regulators below or at the fringe of consciousness. The specialized organs demand an equally specialized biochemical composition in order to achieve their tasks. Most of this specialized biochemistry is concerned with methods for the conversion and transmission of energy and information. The sensory receptors have to convert information arriving as light, heat, or sound waves into forms which can be transmitted through the nerves to the brain. The brain has to process this information so as to arrive at two objectives: one an immediate set of instructions to be dispatched to the muscles, the other a permanent or memory store which will be of subsequent use in helping determine later actions. Finally, the muscles have to recognise and respond to the information arriving at them from the brain by contraction or relaxation: in contracting they have to perform mechanical work, and, to do so, they need a source of energy.

By now, we should have no difficulty in framing the biochemist's

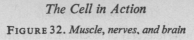

FIGURE 32. *Muscle, nerves, and brain*

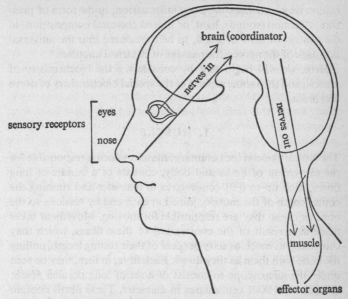

brain (coordinator)

nerves in

nerves out

eyes

sensory receptors

nose

muscle

effector organs

question: what special chemical characteristics can we observe in these organs to account for their physiological role? This question has indeed existed for as long as biochemistry itself, and is, at one stage or another, even if it must be confined to their off-duty moments, on the lips of every biochemist. Ultimately, we only recognize life by its power to act upon the inanimate world around it, and our biochemical analysis cannot stop short merely at showing how the body, left to itself at a suspended interval of time, is composed and functions as an internally self-regulating homeostat; in the long run, we must also hunt for the biochemical mainspring of the body's relationship with the external world.

A good deal is in fact known of the actual workings, biochemical as well as physiological, of nerve, muscle, and brain – much more than we have space to consider here. We must reluctantly leave out, for example, any description of the biochemistry of the special receptor cells of sense organs, although a study of each would reveal an exquisite interplay of biochemical and structural specialization which enables incoming information, in the form of pressure, vibration (sound), light, or indeed chemical composition, in the context of taste or smell, to be transduced into the universal language of the nerves, the passage of electrical impulses.*

Here, we will begin first by considering the biochemistry of muscle, and then some aspects of the special biochemistry of nerve and brain.

I. MUSCLE

The typical skeletal (or voluntary) muscle, which is responsible for the movement of limbs and body, consists of a bundle of long fibres, each up to 0·01 centimetres in diameter and running the entire length of the muscle, joined at each end by tendons to the bone or organ they are responsible for moving. Movement takes place as a result of the contraction of these fibres, which may shorten by as much as sixty per cent of their resting length, pulling the bone with them as they do so. Each fibre, in turn, may be seen under the microscope to consist of a set of long parallel *fibrils*, each about 0·0001 centimetres in diameter. These fibrils contain the contractile material of the muscle. Like other cells, they also have a nucleus, mitochondria, microsomes, and cell membrane, but when examining them under the microscope one is scarcely conscious of these; one sees the fibril primarily as a long, thin, striped tube. These *stripes* are the most remarkable thing about the skeletal muscle fibril; they run across it at right angles at regular intervals, like the black and white markings on the pole of a belisha beacon. What is more, the stripes of adjacent fibrils all run in parallel, so that the stripes appear to continue right the way across the entire muscle fibre. It is this striped quality that particularly characterizes skeletal muscle, as opposed to certain types of smooth muscle which are not normally under voluntary control, but are responsible, for instance, for intestinal or stomach movements.

* It would be less than human of me if I did not refer those interested in this theme to my book *The Conscious Brain* (Penguin, 1976) for further exciting details!

The Cell in Action

Viewed under a higher powered microscope, the stripes may be resolved into alternate light and dark bands, forming the complex but regular patterns of Figure 33. The microscopists refer to these

FIGURE 33. *The muscle*

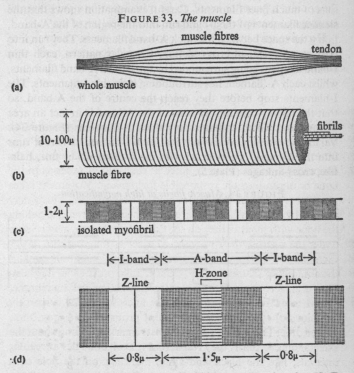

(a) whole muscle

(b) muscle fibre

(c) isolated myofibril

(d)

Muscle at different magnification. (Rabbit psoas muscle, after H. E. Huxley) ($1\mu = 10^{-4}$ cm.)

bands as the I- (light) and A- (dark) bands, the H-zone (lighter area in the middle of a dark A-band) and Z-line. The one A-band, two half I-bands, and one H-zone between each pair of Z-lines represents a single unit, called a *sarcomere*. Each single unit is no more than 0·0002 centimetres in length, so to each fibril there are several thousand sarcomeres joined end to end. When the magnification is increased still further, using the electron microscope, the A- and I-bands break down into a set of filaments running parallel to the axis of the fibril. The A-band is dark because the filaments within it are thick; the I-band, on the other hand, consists of much finer filaments. Careful examination shows that the slender filaments of the I-band run into the region of the A-band, filling the space between the thick A-band filaments. They run into it so as to form, in cross section, a regular pattern, each thin I-band filament being surrounded by three thick A-band filaments, whilst each A-filament has surrounding it six thin I-filaments. The I-filaments stop before they reach the centre of the A-band, so that the lighter-coloured H-zone of the A-band is in fact an area of the band which contains only thick A-filaments (Figure 34). Very close examination reveals that each I-filament, where it runs into the A-band, is connected with the A-filaments by fine, hair-like, cross-linkages (Plate 5).

FIGURE 34. *Muscle fibrils at high magnification*

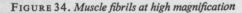

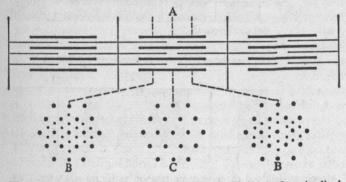

The arrangement of the filaments in the muscle fibrils. A: Longitudinal view. B and C: Cross sections showing thick and thin filaments. (After H. E. Huxley.)

Now when contraction of the muscle occurs, the sequence of events seen under the microscope is that the filaments themselves do *not* change length, but that the thin I-filaments slide in between the thick A-filaments until all the H-zone is filled up; when this happens the normal limit of contraction is reached. Thus the effect is that the I-bands seem to disappear whilst the A-bands remain unaltered, and the two Z-lines defining the limit of the sarcomeres are drawn in towards the central region.

The mechanism of muscular contraction, then, seems to demand the existence of a set of two types of interlocking filaments, contraction occurring when the filaments are slid together, relaxation as they move farther apart. In essence, muscular contraction is more like the shutting of a telescope than the tautening of an elastic band.

What can the biochemist add to this microscopist's and physiologist's view of muscle? It was early recognized that the contractile elements of the fibre were almost entirely composed of protein. At first, it was believed that only one type of protein was present, and it was called *myosin*. It is now known that there are at least four components, myosin, actin, tropomyosin, and troponin. Myosin is the largest of the proteins, with an estimated molecular weight of 460,000 and constituting the greatest proportion (about fifty-four per cent) of the total protein. Actin constitutes about twenty per cent of the total protein but it is a much smaller molecule with a molecular weight of only 41,000. X-ray diffraction patterns for myosin show it to be a fibrous α-protein; that is, it is composed of two polypetide chains twisted in a helix for ninety per cent of its length. In addition myosin can be split by enzymic digestion into two subfractions known as light and heavy meromyosin. Actin on the other hand is a globular protein.

The proteins can be separately extracted from muscle, and both types of molecule can be shown to be long and fibrous in shape. In addition, myosin has the remarkable property of being capable of being extruded through a nozzle to form long artificial fibres (a process analogous to the formation of artificial silks such as rayon). When actin and myosin are extruded together, they form a complex fibre, actomyosin, and it was found by F. Straub and A. Szent-Györgyi in Hungary in 1942 that, if this actomyosin fibre is placed in a solution containing ATP and suitable salts, it will *contract* spontaneously, simultaneously dephosphorylating the ATP. Neither myosin nor actin by themselves, however, would behave like this.

Now it had already been known for some years that ATP and creatine phosphate were broken down during extended muscular contraction to release ADP and inorganic phosphate. Indeed it was these observations, made with muscle in the early 1930s, that first led to the recognition of the role of organic phosphates in biochemical reactions. In 1933, V. A. Engelhardt in Moscow was able to show that, in the presence of myosin, ATP was rapidly split to ADP and inorganic phosphate. Thus myosin seemed to function as an ATP-ase enzyme. It was clear that these observations must be related in some way to the explanation of muscular contraction. The situation was further complicated though by the observation that the injection of a tiny amount of Ca^{++} ions into a single muscle *in vivo* will produce a local contraction although the purified actomyosin complex does not require Ca^{++} for its contraction.

However, it required the coming of the electron microscope and the techniques of X-ray crystallography to be able to link decisively the biochemical and microscopical observations, and these later interpretations were to a large extent the work of H. E. and A. F. Huxley and their co-workers in England. According to the present conception (at least in its simplest form), the two major proteins, myosin and actin, correspond to the two different types of filament in the muscle fibril. The thick filaments of the A-band are composed of *myosin*, the thin I-band filaments are *actin*. The myosin molecules in the thick filaments form rods composed of two chains. Each chain has a thickened portion at one end giving the rod a double-headed appearance. It is these 'heads' that come into contact with the actin molecule and they are composed of heavy meromyosin which is the ATP-ase carrier. The direction of the meromyosin heads reverses in the centre of the thick filament so that each filament is bipolar with a central bare zone, visible under very high magnification, containing no meromyosin heads. As we have said, actin is a globular protein and the thin filaments are formed by two strings or chains of actin molecules wound loosely together like two strings of beads. It is only recently that the association of troponin and tropomyosin with the actin filaments has been clarified. Tropomyosin is a fibrous protein which seems to lie in the groove formed by the winding of the two actin chains. Troponin is arranged at regular intervals along the tropomyosin molecule.

With this structural information, we are now ready to consider the contractile process, at least as far as current understanding will allow us. The signal for contraction in the type of muscle we have been considering is the release of a chemical transmitter from nerve endings at the junction between nerve and muscle, which we will shortly consider in more detail. This leads to a depolarization of the muscle cell membrane, which in turn results in the release of a large number of Ca^{++} ions into the muscle fibre – the amplification process. It is these Ca^{++} ions that act as the second messenger and react with the effectors, that is, the muscle proteins, to bring about contraction. Thus the initial stimulus has been coupled to the contractile process by the Ca^{++}. It has been suggested that Ca^{++} combines with the troponin and this in some way causes the tropomyosin in the same region to shift slightly, exposing a previously hidden part of the actin molecule.

Now both the actin and myosin molecules carry strongly negative charges on their surfaces – indeed it is this electrochemical repulsion that helps maintain the structural integrity of the whole muscle. The minimum potential, that is, the most positive region, exists at the mid point between the two filaments and it is this region that is occupied by the meromyosin projections containing the multivalent and electronegative ATP system. The effect of Ca^{++} binding to the actin filament will be to make the surface of the filament temporarily electropositive over a very localized area. The ATP complex will be attracted towards it and will interact. The ATP is then broken down to ADP with the resultant release of energy. Cessation of the nervous impulse causes an immediate reuptake of the calcium ions from the fibrils, the meromyosin/ATP complex is released back into the inter-filament space and the tropomyosin/troponin component shifts back into the actin groove, but in a slightly different position, relative to its original alignment with the myosin filament. Thus the next site of attachment of the meromyosin molecule will be to a troponin molecule farther along on the actin filament, and the process is repeated. During contraction then the muscle shortens *not* by any change in length of the filaments but by the filaments sliding past each other, gradually filling the H zone. The process is summarized in Figure 35.

FIGURE 35. *Structure of myofilaments and stimulus contraction coupling*

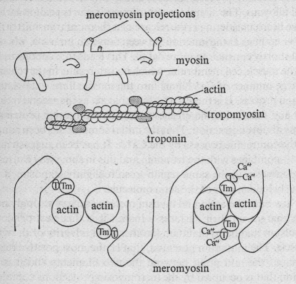

This model of muscle action incorporates many of the observable facts, but there are also many more questions that remain to be answered, particularly the exact involvement of Ca^{++}. Nevertheless muscle cells provide a prime example of specialization of structure to fulfil a specific role – to enable the organism to respond quickly and efficiently to a changing external environment.

II. NERVE AND BRAIN

The great mass of cells of the central nervous system is concerned with the co-ordination of the millions of incoming messages arriving by the second from the internal and external sense organs of the peripheral nervous system, the control of the motor activities of the myriad of muscles, and the continued matching and comparing of the messages and responses which represent the present moment of existence of the organism with the accumulated wisdom of its past history, relating this to present needs and possible outcomes of actions. Of all the things that might be said of the biochemistry of nerve and brain at this point, we will confine ourselves here, by a process of tunnel vision, to just two. We wish to describe how the structures of the nerve cells are specialized to enable them to perform their functions, and what the biochemist can say about two fundamental properties of cell nerve cells, their capacity to conduct messages (impulses) over relatively long distances, and to communicate the import of these messages either to other nerve cells, or, in the last analysis, to effector organs such as the muscle. Like the muscle itself, the nerve trunk is made up of a number of smaller *nerve fibres*; in what follows, we will use the word nerve to apply to one single such fibre. The nerve fibre, then, is basically a long tube which connects the top end of the nerve cell, generally located in the brain or spinal cord, with the muscle. Wrapped round this tube, or *axon*, which is in fact part of a long-drawn-out single cell, is a layer of fatty material (*myelin*), which provides it with a protective sheath, in appearance exactly like the insulating rubber round the metal core of an electric cable, and, as we shall see, serving much the same function.

At the top of the nerve axon is a large, diamond-shaped swelling which is the *nerve cell body* (*neuron*), which contains all the conventional apparatus of the cell; nucleus, mitochondria, and microsomes. The cell body is so large in comparison with the axon that it sits at the top of it not unlike a toffee-apple at the end of its stick. Where the axon, at its lower end, comes into contact with the muscle cell, it again spreads out slightly, so as to form a sort of broad foot which treads on the muscle cell, and, as a result, touches it over a quite considerable area.

This region of contact between the two cells is called the neuromuscular junction. The neurophysiologist's name for such a junction is *synapse*. If, as frequently happens, the distance between the brain or spinal cord and the muscle is too long for one axon to stretch the whole way, a break is made half way between them. The nerve leading from the brain then forms a synapse with the cell body of a second nerve, whose axon can then reach the remaining distance to the muscle. Such a synapse between two nerve cells, as well as being a means of sending a signal over long distances, also, as we shall see, provides a point at which the message of command, descending from the brain, can be modified to take additional factors into consideration. The message coming from the brain crosses the synapse between the first nerve and the second and continues down the second nerve to the neuromuscular synapse where it causes the muscle to contract. In general, whenever it becomes necessary to transmit a message from one part of the nervous system to another, it is done by means of such a synapse. A similar system carries messages *from* sensory organs such as eyes and ears *to* the brain, and yet further synapses interconnect the large number of cells within the brain itself. Thus messages can be rapidly transmitted over considerable distances and to a large number of different nerve cells or muscles (Figure 36).

FIGURE 36.

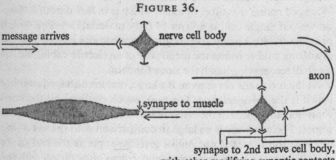

The highest region of the brain, the cerebral cortex, contains an astronomic number of such nerve cells (something like 10,000,000,000 in humans) each one of which may make ten thousand or more synaptic contacts with others. The richness of such interconnections accounts for the tremendous ability of the brain to co-ordinate, compare, and control the activities of the sense organs and muscles, and to learn from experience. *En masse*, this huge array of interconnecting nerve cells within the brain acquires characteristics which transcend the behaviour of any individual cell, just as the activity of a computer transcends the properties of any one of its transistors or magnetic tapes.

The essential functions of the nerve cell are three. At the head of the toffee-apple, the cell-body has to receive and respond to the incoming messages arriving from other nerve cells which synapse with it. These messages must be passed swiftly down the axon, and finally, at the bottom, must be transmitted across the cell membrane to the nerve or muscle cell at the other side of the synapse. Over the last ten years, a good deal of the biochemical mechanism of each of these three steps has been unravelled.

The simplest part of the process to understand is the way the messages pass down the nerve axon itself. Here our earlier description of the nerve fibre, surrounded by its fatty sheath like an insulated electric cable, comes strikingly into its own. In fact, messages pass down the nerve axon in the form of electric currents which begin where the axon springs out of the nerve cell body and travel down the fibre at a rate of about twenty metres a second.

Two sorts of job are done by more familiar electric currents. A steady current flowing to an electric light bulb provides a constant source of *power* for the bulb; if the flow of electricity alters, the bulb flickers. A door-bell, on the other hand, must ring briefly and then stop. A short burst of electricity is wanted here, to act as a *signal*. The currents flowing in the nerve fibre are signals, not power supplies. Thus if we record the passage of electricity down a nerve we find not a steady reading, of say four volts, such as one might get from a torch battery, but instead, at any one point along the nerve, the voltage suddenly fluctuates, rises to a peak, and declines again as the electric signal arrives, passes the recording point, and is gone. A *wave* of electricity has passed down the fibre. If we draw a curve plotting the rise and fall of voltage at a particular point on a nerve fibre during the passage of an impulse down it, we get the picture shown in Figure 37.

What is the mechanism of this remarkable phenomenon? In the sort of electricity produced by a generator or battery, the current running through the cables is in fact carried by a *flow of electrons* moving down the wire from the negative to the positive terminal. In animal-generated electricity, the current is still carried by a flow

FIGURE 37. *Voltage changes as impulse passes down nerve fibre*

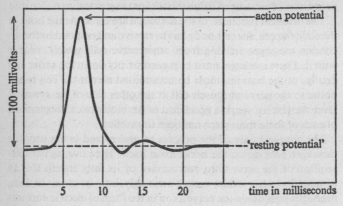

of charged particles, but, instead of the negatively-charged electrons, the carriers are positively-charged ions. That this flow is
possible depends on a fact we have already commented on on
more than one occasion – that the ionic composition of the inside
of a cell is quite different from that of the outside (see page 88, for
instance). Inside, there is a high concentration of potassium, but
little sodium. Outside, there is a great deal of sodium but only a
small amount of potassium. Nerves resemble most other cells in
this. Sodium and potassium, as we know, exist in solution as their
charged ions – each carries one positive charge, and we have become accustomed to writing them Na^+, K^+. Of course, the cell
contains many other charged ions as well: magnesium and calcium (Ma^{++}, Ca^{++}) amongst the positive ions, and chloride
and phosphate amongst the negative ions (Cl^-, HPO_4^{--}). In
general, there are about as many negative as positively charged
ions within the cell, so that the overall *net* charge is zero, and the
same is also true of the fluid in which the cell is bathed.

But the fact that the concentrations of sodium and potassium
are different either side of the cell membrane results in certain
peculiarities despite this overall electrical neutrality. For, in a
non-living system, we would expect that if we separated a solution
of potassium chloride from one of sodium chloride by a thin,
permeable membrane – say a cellophane sheet – potassium ions

would pass through the barrier in one direction, and sodium in the other, until the concentration of both ions was equal on either side of the membrane. That this does not happen in living cells is because the cell membrane behaves, as we discussed in relation to the general properties of the cell membrane in Chapter 4, as if it were impermeable to sodium. It permits potassium to enter the cell but not to leave, but does not let sodium enter at all. This is not entirely a passive refusal of admittance, for if an experiment is made in which sodium is injected into the cell, it is quickly pumped out through the cell membrane again *against* the sodium concentration gradient. The membrane is thus a *dynamic* system capable of selectively distinguishing between sodium and potassium ions.

What is the effect of these differences in concentration? If potassium is not to flow out of the cell down a concentration gradient, it must be in response to some other force acting inwards. This force is provided by the positively-charged sodium ions, which line up on the outside of the membrane so as to provide a barrier of positive charge which repels the positively-charged potassium ion. At the same time the sodium ions tend to attract *negatively*-charged ions such as phosphate or chloride across the cell membrane. The net effect is that we can draw the axon as shown in Figure 38a with a line of positive charges down its outside surface and of negative charges on its inside. A potential difference thus exists between the two sides of the membrane – if they were shorted by a wire running between them they would register a voltage. This can, in fact, be done, if a tiny glass microelectrode is inserted into the axon, and connected through a voltage recorder (galvanometer) to the outside surface of the nerve. It is also possible to calculate a theoretical voltage which should arise in a system in which a membrane permeable to potassium but not sodium exists – a comparatively simple piece of mathematics. Fortunately, the theoretical and experimental results agree; the voltage recorder indicates that the inside of the cell maintains a potential sixty-five to ninety-five millivolts negative to the external surface. The existence of such a voltage is absolutely identical in principle to the voltage produced by the dry chemical battery of a torch or the wet battery of an accumulator.

Now consider what happens if for some reason at one point

FIGURE 38. *Nervous transmission*

(a) At rest, charged ions at either side of nerve membrane provide a potential difference.

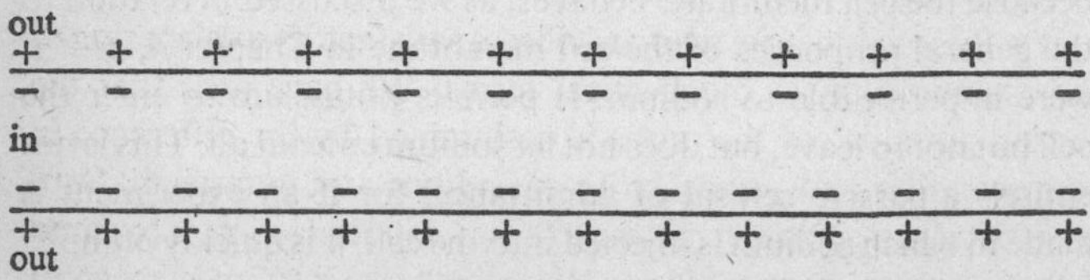

out
+ + + + + + + + + + + +
− − − − − − − − − − − −
in
− − − − − − − − − − − −
+ + + + + + + + + + + +
out

(b) A local depolarization at A results in current flowing, and the message passes down the nerve fibre.

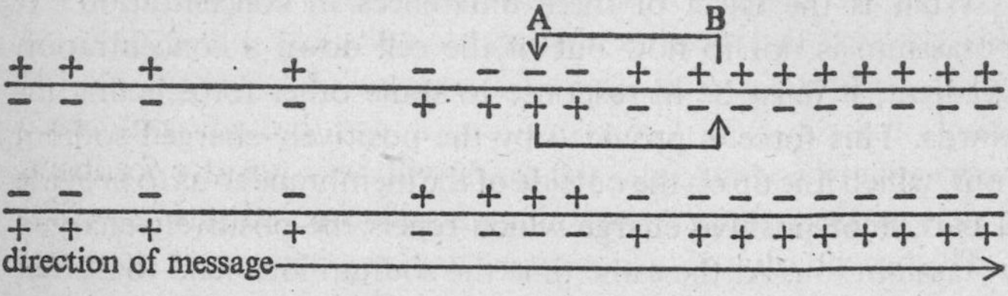

direction of message ⟶

along the nerve axon the set of charges is reduced to zero, for example, by the application of an electric shock. This condition is referred to as the *depolarization* of the nerve, and its results are indicated in the diagram Figure 38(b). The results of depolarization are to cause at a certain point on the nerve a reversal of the set of plus and minus charges within and without the membrane. Thus, at the outside of the membrane, one region, A, becomes negative with respect to a region farther along, B, and, correspondingly, within the membrane B becomes negative with respect to A. A tiny *local circuit* is completed between A and B and *current flows between them*. But the effect of the arrival of the current at B is to depolarize B in its turn, and hence render it negative to a point still farther along the axon; another local current then begins to flow and the depolarization is repeated. Thus a depolarization at one end of the nerve axon results in the establishment of a series of local circuits, which, moving like a wave over the surface of the axon, will carry the depolarization along its length. Provided the initial stimulus is large enough to set the circuits going, it is carried rapidly down the nerve fibre. This method of propagation of nerve impulses, suspected for many years, was finally verified experimentally by A. L. Hodgkin, R. D. Keynes, and their collaborators in Cambridge.

It was Hodgkin, too, who provided a satisfactory explanation of the phenomenon in biochemical terms. We have seen that the resting potential of the nerve is maintained by the difference in concentration and permeability of sodium and potassium. Suppose, Hodgkin argued, that the effect of depolarization is to cause a change in the permeability of the membrane so that it becomes for a brief while very permeable to sodium, what will happen? Sodium will flow rapidly into the nerve axon down the concentration gradient, entering faster than potassium can leave. As this happens the membrane potential will drop still further, as positive charges are transferred from outside to within. As the membrane potential drops further, it will become still more permeable to sodium, and the result will be a positive feedback system in which the entry of sodium is self-stimulating (Figure 39).

This inrush of sodium continues until the membrane potential of eighty to ninety millivolts negative is reduced to zero and finally converted to one of some twenty to thirty millivolts positive. This represents the 'spike' of the wave of Figure 37. Sodium entry now ceases, as it is having to enter *uphill* against a potential gradient. At the same time, according to the Hodgkin model, the potassium permeability of the axon increases, and potassium leaks out down *its* concentration gradient until the membrane potential falls to its old value of about eighty-five millivolts negative and the electrical wave has passed on down the axon. In fact, potassium exit goes on slightly longer than is strictly necessary, and the potential slightly overshoots its original value, resulting in the second 'trough' of Figure 37. During this period, where the potential is some milli-

FIGURE 39.

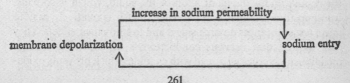

volts below the normal resting level, the nerve becomes incapable of carrying a second message. Within a few thousandths of a second, however, the old value has been restored once more and the nerve is again excitable and ready for action.

Hodgkin's theory has been amply borne out by experiments. Studies using radioactive sodium and potassium have been able to demonstrate the changes in entry and exit of these ions across the membrane during nervous stimulation, whilst it has long been known that, just as the theory would predict, nervous conduction becomes impossible if all the external sodium is removed. For his part in this work, Hodgkin shared the 1963 Nobel Prize for medicine.

The changes in membrane potential are achieved at the expense of the sodium and potassium levels which the cell membrane normally maintains. The changes we have described as occurring at the membrane result only in rapid local changes of concentration, so that they hardly affect the overall differences in sodium and potassium levels inside and outside the cell. Despite the local changes, these concentration differences remain so great that, in a large nerve, upwards of a million impulses could be carried before the decline in internal potassium and rise in sodium became great enough to inactivate it. Nonetheless, ultimately, the nerve must set to work to rectify the changes in concentration and restore the old differentials between sodium and potassium. It may be noted that, until now, processes we have described have not demanded the output of energy by the nerve; the changes in membrane permeability instead utilized the potential energy latent in the differences in sodium and potassium concentration across the membrane. The work of the nerve consists in establishing this potential energy once more.

Again, the critical experiments which demonstrated the mechanisms of this process were performed by Hodgkin, by Keynes, and by A. F. Huxley. They used in their experiments the biggest nerve they could find – the giant axon of the squid, which is so wide, being up to a millimetre in diameter, that it is relatively easy to inject test substances down inside it and to study their effects. They first showed that nervous conduction is dependent on energy metabolism. Nerves poisoned with cyanide, which prevents oxidative phosphorylation, rapidly lose their ability to conduct impulses. But if ATP is injected into the poisoned nerve, it becomes active once more, and can go on conducting until all the ATP is broken down once more. Thus, just as in muscle, the ultimate energy on which the cell draws for its ability to act is the energy of ATP.

Later work has been able to show that the ATP is in fact utilized in order to maintain the sodium and potassium levels of the nerve. ATP is broken down during the extrusion of sodium and the entry of potassium. The entry and exit of these ions are found to be linked processes, and, in general, one molecule of ATP is broken down for about every three ions of sodium and potassium transported across the membrane. The exact mechanism whereby the energy of ATP is utilized in this way is still unknown, but it presumably involves phosphorylation of membrane proteins so as to induce conformational changes in carriers, or open pores through which ions can pass, according to the general principles discussed earlier in relationship to membrane properties, but the full details of this process are even less well-understood than the fine mechanism of the coupling of ATP breakdown to muscular contraction, even though, just as with that problem, one is tantalizingly close to the solution.

Transmission of impulses at synapses

The nerve impulse flows down the axon in a wave of depolarization of the membrane. But at the synapse a different problem arises. The membrane of the nerve cell or muscle fibril at the other side of the junction is not continuous with that of the incoming nerve. A means must be found of bridging the gap between the two cells. Essentially, this demands that the arrival of the depolarizing wave at the end of one nerve axon acts as the trigger or signal for the start of a similar wave in the nerve cell and axon of the second nerve. This triggering action depends on the existence of *chemical transmitters*. The arrival of the nerve impulse at the synapse stimulates the release from the nerve endings of a chemical which can diffuse through the membrane of the first nerve to arrive at the cell body of the second (Figure 40). There, it causes depolarization of the membrane and triggers an impulse in the second nerve. The chemical nature of these transmitter substances differs in different

The Chemistry of Life

FIGURE 40. *Transmission at the synapse*

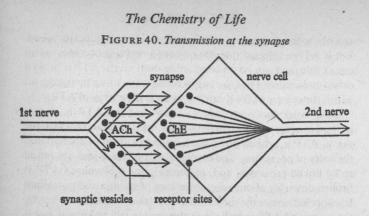

Acetylcholine, released from synaptic vesicles by incoming stimulus, diffuses across synapse to receptor sites in second nerve, where it is destroyed by cholinesterase, depolarizing second nerve in process.

parts of the nervous system. Some are simple amino acids, like glutamate. Others are amines, substances related to the hormone adrenalin described in the previous chapter. Still others – a whole new, and recently discovered class of transmitters – are small peptide molecules. But one of the commonest is the substance acetylcholine. It seems to be the case that different pathways in the brain and nervous system, in which many nerve cells are involved, utilize different transmitters; thus in some of the brain pathways involved in emotional and attentional responses, amines seem to be the transmitters; some of the small peptide transmitters may be involved in the transmission of pain sensations (there was much excitement in the mid-1970s when one of these transmitters was found to be chemically rather similar to the artificial pain-reliever morphine). Many of the synapses between nerve and voluntary muscle are mediated by acetylcholine, and we can look at its mode of action in a little more detail.

Acetylcholine is manufactured by the acetylation of the fat *choline* with acetyl-CoA, and the responsible enzyme, *choline acetylase*, is present in the nerve endings. Both it, and the acetyl-choline itself, are held in a series of particles, in many ways similar to the lysosomes, and which have been called by their discoverers, Victor Whittaker in Cambridge and George Grey in London, *synaptic vesicles.* These particles, distributed through the nerve ending but not found elsewhere, contain the acetylcholine in a bound, inactive form.

Additional acetylcholine, however, is found free in the cytoplasm of the synaptic vesicle. There is a school of thought which maintains that the vesicles are simply reserve stores of the transmitter. However, probably the majority of people working in the area believe that the effect of the arrival of the nerve impulse is to cause many of the vesicles to move towards the membrane, fuse with it and empty their contents into the gap between the two cells, the synaptic cleft, above which they can diffuse to the second nerve cell or to the muscle.

On the postsynaptic side, there are specific receptors located in the membrane on to which the transmitter binds, in a similar way to the type of hormone-receptor interaction proposed in the previous chapter. (It is also possible that cAMP is involved in the postsynaptic response to some transmitters.) The transmitter-receptor results in a change in the postsynaptic membrane structure. If the receptor is an excitatory one, this may result in an influx of Ca^{++} ions large enough for the postsynaptic membrane to become depolarized. If a sufficient number of synapses transmit excitatory messages to the postsynaptic nerve at around the same time, the result will be a general depolarization, and the second nerve will 'fire' or the muscle contract.

On the other hand, there are other transmitters which, working in the same way, produce a hyperpolarization of the postsynaptic nerve (i.e. make it *more*, rather than *less* negative) – thus making it harder to fire. These are *inhibitory* transmitters and are of equal importance to the excitatory ones (nerve cells need to be able to say No as well as Yes!).

Once it has exerted its effect, the transmitter must be destroyed to prevent its presence from interfering with further messages. In the case of acetylcholine, this is achieved by the enzyme *acetylcholinesterase*, which, present in the postsynaptic membranes, hydrolyses it again to choline and acetic acid. Other transmitters are reabsorbed after use, either into the presynaptic cells or into a further class of cell, the glial cells, which are present in nervous tissue in large numbers, surrounding most of the nerve cells and the synapses. (Some of the work of my own laboratory in the last decade has been concerned with the relationship between the biochemistry of neurons and glial cells and of these complex questions of factors determining nerve cell transmission.) The whole process of synaptic transmission is summarized diagrammatically in Figure 40.

The neurotransmitters like acetylcholine establish contact between two cells across the gap which separates them. It is not surprising to find that this is a very vulnerable link in the chain of command between nerve and muscle. Anything which interferes with the diffusion of the transmitter across this space will prevent the message being passed on, and many drugs which interfere either with transmitters, the receptors, or their subsequent destruction by postsynaptic enzymes, thus block synaptic transmission and result in a sort of chemical paralysis. For acetylcholine, *nicotine* is one such drug; another is the poison which Central American Indians used to dip their arrows into and which has since become beloved of crime-writers, the mysterious *curare*. In more recent times another class of substances which affect cholinesterase have achieved more sinister significance, though. Originally a product of German wartime research, a group of organophosphorus compounds which act in this way were developed as the nerve gases. An even more toxic variant (a milligram or so absorbed through the skin is said to be lethal) was developed in the mid 1950s at Porton, Wiltshire, the so-called V-agents. They are amongst the most poisonous chemicals known. If the internal working of our cells depends ultimately on ATP, it may be maintained that our ability to control and command our body to act is similarly dependent on the neurotransmitters.

Many other specialized drugs and pharmacological agents interfere with transmitter substances. Some of those used in the control of mental illness, antidepressants and tranquillizers, work this way, and indeed there are proposals that some of the mood-affecting psychedelic agents, like cannabis (marijuana) and LSD, exert their effect by interacting in some subtle way with particular central nervous system transmitters. There is no doubt that the area of brain biochemistry has been amongst the most fertile of recent years, and will be of increasing excitement in the future.

CHAPTER 13

THE UNITY OF BIOCHEMISTRY

In what has gone before, we have tried to describe the behaviour, in chemical and physical terms, of the biological unit called 'the cell'. But when speaking of this cell, we have in fact nearly always been referring to, not the cell of any one of the millions of different species of living things that inhabit the earth, but the cell of one particular group of animals – the mammals. Although on rare occasions we have digressed into the fields of plant and microbial biochemistry, such trespasses have nearly always been brief, and then normally intended to illustrate a point about the mammalian cell.

Indeed, we are not even being very fair in extending ourselves as far as even a description of 'the mammalian cell'. Most laboratory work, and hence the experiments we have been describing, is done on only a very small number of different species. Rats and guinea-pigs, rabbits and hamsters are studied by biochemists in their tens of thousands; dogs, cats, and domestic animals such as sheep, cows, and pigs have also provided starting material for research. Occasionally experiments are made with birds such as chickens and pigeons. Always, attempts are made to relate this biochemistry of laboratory animals back to humans. But these species form only a very small fraction of the widely differing groups of known mammals. And when we come to plants, insects, or micro-organisms, the number of whose known species overwhelmingly outnumbers the total of the mammalian kingdom, our biochemical studies seem to fade into insignificance.

Yet we have confidently described 'the cell', its enzymes, metabolism, and behaviour almost as if unaware of the existence of so many other different sorts of cell remaining to be examined. Have we not been grotesquely over-confident in doing so?

Fortunately, we are fairly secure in saying no. The differences in appearance and overall behaviour between humans and yeast are indeed vast. Yet it is astonishingly, almost at first sight unbelievably, the case that most of the chemicals that compose the

two are practically identical; almost without exception, their proteins are made of the same twenty amino acids, their nucleic acids
of the same four purine and pyrimidine bases, their carbohydrates
of the same or similar sugars. And even where slight differences
may be pinned down between, say, the amino acid sequences of
individual enzymes, the metabolic pathways catalysed by these
same enzymes remain, in yeast and humans, identical over large
regions. The pathway of glucose breakdown, studied originally in
yeast as the fermentation of sugar to alcohol, was subsequently
found to be identical with the route by which the muscle cell converted glucose to pyruvic acid. The only difference lay in the subsequent fate of the pyruvic acid.

Similarly, it is possible to purify an enzyme from one organism
and use it without apparent difficulty to study a reaction in another
quite different one. Ribosomes from a micro-organism and soluble
fraction from a rabbit or duck will contentedly collaborate to
synthesize protein. Even though the biochemical behaviour of only
a minute percentage of the different forms of life on earth have
been examined, one may still predict with a good chance of success
that the general conclusions may be extrapolated to cover all other
forms as well. The basic mechanisms of carbohydrate, fat, and
protein catabolism and synthesis are the same in all forms of life
now existing. This is a fact at first sight so unexpected and so
surprising when one thinks of the manifold differences of form
which life manifests, that it demands almost a positive effort of
will to accept it.

Its implications become all the more startling when one comes
to consider the undoubted biochemical differences that do exist
between different life forms. The most profound of these differences lie in the sources from which the organism obtains the
energy it requires in order to remain alive. Animals, fungi, viruses,
and most bacteria rely on the existence of preformed organic compounds. We have dwelt at some length on the way in which
animals burn glucose to carbon dioxide and use the energy so
released to synthesize ATP. Deprived of glucose, or its rather in-
effective substitutes in fat and protein, an animal rapidly wastes
away and dies. In the presence of glucose and certain essential
amino acids and vitamins, it can synthesize all the thousands of

other chemicals it requires. Many bacteria are less demanding; they can survive on simpler 2- or 3-carbon organic acids, and some of them do not need amino acids but can make their own by transamination reactions with ammonia. Yet even these rather cleverer bacteria and fungi can normally exist as well (or better) in the presence of glucose and amino acids as they do in their absence. Most of them have the enzymic ability to deal with these substances just as animals can, even if the bacteria *can* do without if times get hard. They are just rather less specialized than the animals, and can therefore live rougher.

A second major difference lies between those organisms which require oxygen to act as an 'energy-sink' and oxidize their foodstuffs, and those 'anaerobic' bacteria which either never use oxygen, or can do without it at a pinch. Yet even such differences are more apparent than real. Those micro-organisms which obtain their energy entirely by fermentation do so by pathways of metabolism similar to the routes taken in animals for the initial steps of glucose breakdown. The difference lies only in the fact that the fermentative organisms are unable to complete the process and oxidize the end-products of fermentation, and instead resort to a variety of tricks to extort the maximum energy from their excreta before finally discarding them as alcohol, acetaldehyde, lactic acid, or other essentially half-digested substances. In this case the animals possess oxidative abilities that the fermentative micro-organisms do not. But their basic biochemistry is not greatly different, only more efficient.

A different class of anaerobes are those which can do without oxygen, not because they do not oxidize their substrates, but because they find an alternative hydrogen acceptor and ultimate 'energy-sink'. Typically, they use either sulphate or nitrate as their acceptors; the nitrate-reducing bacteria, for example, convert nitric acid, HNO_3, to nitrous acid, HNO_2, and in doing so gain an atom of oxygen to which they can pass hydrogens and reduce to water. But the most fascinating thing about these nitrate-reducing bacteria, seemingly operating on such different principles from the animal world, is that the electron-transport pathways by which they pass their hydrogen to nitric acid are identical with those of the animals which use the oxygen of the air. The substrate yields its hydrogen to a dehydrogenase which passes it on to the intermediate hydrogen carriers which are cytochromes. The only difference is that, in the final stage, instead of using cytochrome oxidase to take the hydrogen from the cytochrome to oxygen, the nitrate-reducing bacteria oxidize their cytochromes with a different enzyme, nitrate reductase, which converts nitric to nitrous acid and water (Figure 41).

FIGURE 41.

The difference between humans and nitrate-reducing bacteria

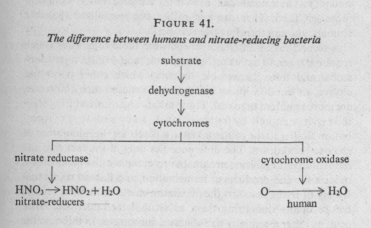

AUTOTROPHES AND HETEROTROPHES

But the real dividing line in the biochemical world comes between those groups we have just discussed, all of which depend on the existence of preformed organic compounds for survival, and those living things which can make all their own organic substances from simple inorganic materials. Such organisms are called *autotrophes*, indicating that they are self-sufficient by comparison with the *heterotrophes*, like yeast and humans, who have to be cushioned by the existence of sugars and amino acids against the harsh realities of the inorganic world.

Autotrophes do not obtain their energy by burning ready-made fuel, but cast about to find an alternative energy-source. The paradigm case, of course, is the green plant, which avoids the heterotrophe's dilemma by trapping the light energy pouring down on to the earth from the sun, and using it to 'fix' carbon dioxide as organic carbon and, ultimately, to synthesize sugars and starches, a process known as *photosynthesis*. The 'higher' green plant is not alone in performing photosynthesis; both algae and a group of photosynthetic bacteria can perform similar reactions. Another group of autotrophes (the *chemo-autotrophes* as opposed to 'photo-autotrophes') use the energy latent in certain inorganic chemicals for carbon dioxide fixation instead – by the oxidation, for example, of sulphur, ammonia, or hydrogen. Although biochemically fascinating, these chemo-autotrophes seem to form an evolutionary backwater, an essay in chemical versatility that did not quite come off, and most attention has, not unnaturally, been focused on the photosynthesizing organisms, particularly as it was at one time believed that their self-sufficiency might provide the clue to the origin of life.

THE MECHANISM OF PHOTOSYNTHESIS

The problems faced by the plant are in essence identical to those of the animal. They may be summed up as the need to trap energy and obtain a source of primary building blocks so they may carry out the biochemical synthesis of more complex molecules. Both forms of life obtain their precursors from essentially the same source – breakdown of glucose to CO_2 and H_2O via glycolysis and the citric acid cycle – the routes and the enzymes used are identical. The real difference lies, as we shall see, in the source of the glucose. Both plants and animals use ATP as their energy reservoir and in both it is formed by linking its synthesis to the passage of electrons along a chain of carriers. Again, the real difference lies in the source of the energy that is trapped in this way. Animals obtain this energy from glucose, and we have gone into some detail concerning the mechanism of ATP synthesis linked to the respiratory process. However, in green plants, in the light, glucose breakdown is not the main source of ATP. Any ATP formed is a result of substrate level phosphorylation at particular stages during the glycolytic and citric acid cycle pathway. Nevertheless, given glucose and a nitrogen source, plants can live for a very long time in the dark, though they are clearly much happier in the presence of sunlight. So even an autotrophic organism will live like a heterotrophe given the right circumstances.

Plants obtain the energy which is trapped into the ATP molecule by transforming the light they receive from the sun. During this process electrons are passed from H_2O to NADP, releasing O_2 and giving reducing power in the form of $NADPH_2$. So the net result is a reversal of the energy conservation process in animals, where $NADH_2$ is oxidized to NAD and O_2 is reduced to H_2O. But the essential feature is that a stepwise electron transfer pathway is again involved and is again linked to ATP formation. It is the ATP and $NADPH_2$ synthesized during this light reaction that is subsequently used to make glucose from the simple inorganic substances CO_2 and H_2O, and it is this set of reactions which make up the process of photosynthesis. CO_2 fixation has been called the dark reaction because, if the plant already has adequate supplies of ATP and $NADPH_2$, it can indeed take place in the dark, but normally, during daylight hours, both reactions take place together. The overall reaction then can be written:

$$6CO_2 + 6H_2O + \text{light energy} \rightarrow C_6H_{12}O_6 + 6O_2 \quad (1)$$

In this way plants provide themselves, and the heterotrophic world as well, with the only ultimate source of organic compounds at present available in the world outside the chemist's synthetic test tubes. The total 'fixing' of CO_2 that occurs by photosynthesis is prodigious, providing as much as 100,000,000,000 tons of organic carbon a year. At the same time it continually renews the oxygen of the atmosphere and removes the carbon dioxide accumulated during respiration, providing a turnover so rapid that every molecule of CO_2 in the atmosphere gets incorporated into glucose by photosynthesis once every 200 years, and every oxygen molecule every 2,000 years or so. Between them, the photosynthetic mechanisms of the plant and the respiratory system of heterotrophes provide for the regular revolution of the 'carbon cycle' which takes so prominent a place in every child's biology book. The analogies between the photosynthetic light reaction and hydrogen transport in animals are not merely chemical or mechanistic, for photosynthesis takes place in a subcellular structure called the chloroplast which has a striking structural similarity to the mitochondrion (Figure 42). Like the cristae of the mitochondrion, the grana and lamellae of the chloroplast provide the sites for ATP synthesis and hydrogen (electron) transport, and have a

FIGURE 42. *A chloroplast*

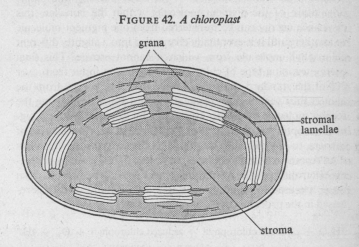

grana

stromal lamellae

stroma

lipoprotein structure similar to that already described for the mitochondrion.

In the trapping of light energy, the first and most critical of the steps of photosynthesis, the substance *chlorophyll* (which gives plants, and, according to the advertisers, some toothpastes, their green colour) is all important. Although chlorophyll is by no means the only photosynthetic pigment, it is the only essential one. The molecule has a hydrophobic hydrocarbon tail by which it becomes firmly embedded in the chloroplast lamellae. The polar head part of the molecule is in fact very similar in design to that of the haem of the cytochromes and haemoglobin (see page 135). Like haem, it consists of a linked series of four carbon-and-nitrogen containing rings ('pyrrolle rings') joined together to form a sort of doughnut with a hole in the middle. This hole is filled in haem by the metal *iron*; in chlorophyll on the other hand

the jam in the doughnut is made of *magnesium*. The ring structures contain a series of alternating double and single bonds, and
the absorption of a given small amount of light (a quantum) of
a particular wave-length causes a sort of vibration, or resonance
around these bonds. Because of the close packing and stable
orientation of the pigment molecules within the lamellae, this
resonance energy can be transferred from one pigment molecule
to another until it is eventually channelled into a slightly different
chlorophyll molecule from which it cannot escape. This final
energy trapping type of chlorophyll receives an input from over
300 of the standard chlorophyll molecules. The energy from the
light is thus very highly concentrated at a single site, giving the
second molecule the ability to transfer an electron to a non-pig-
ment receptor which in turn passes it, via an intermediate set of
carriers, to NADP. The chlorophyll at the central site, which has
thus become oxidized, is converted to its original state by accepting electrons from the hydroxyl group of water. The protons from
water are used to reduce NADP to $NADPH_2$ and oxygen is released in the process:

$$2H_2O + \text{oxidized chlorophyll} \rightarrow \text{reduced chlorophyll} + O_2 + 4H^+$$

$$2(H) + NADP \rightarrow NADPH_2$$

Recent research work has shown that there are in fact two
different such reaction centres each receiving energy from two
different pigment systems. The situation is further complicated by
the fact that flow of electrons from H_2O to $NADPH_2$ involves the
co-operation of both pigment systems in a sequence of events
worked out by Hill and Bendall in 1960 and known as the Z
scheme. The Z scheme bears, once again, a remarkable similarity
to the respiratory electron transport chain, although it is even
more complex and still poorly understood in parts, particularly the
identity of many of the electron carriers. It is known, however,
that they include plastoquinone, two cytochromes (f and b) and
ferredoxin. The photosynthetic phosphorylation of ADP to ATP
occurs if the difference in redox potential between the electron
carriers is sufficient to allow this endothermic reaction to take

place, as with oxidative phosphorylation in the mitochondrion, but the exact mechanism by which this is achieved is still being investigated. Given the other similarities with respiratory phosphorylation the mechanism may well involve the pumping of protons across membranes as in chemiosmosis (page 138) and indeed in some of the original demonstrations of the existence of trasmembrane, proton gradients were carried out with chloroplasts.

Let us return to the so-called dark reaction of photosynthesis. In it, both the $NADPH_2$ and the ATP formed in the light reactions are consumed in the fixation of CO_2. The fixation reactions were charted by Melvin Calvin and his co-workers in Berkeley, California (for which Calvin received the Nobel Prize for 1961), with the use of radioactive carbon dioxide. During these reactions, CO_2 is made to combine with a pentose (5-carbon) sugar, ribulose diphosphate, to give an unstable 6-carbon intermediate which breaks down to two molecules of the 3-carbon phosphoglyceric acid.

$$C_5 \quad + \quad C_1 \quad \rightarrow \quad C_6 \quad \rightarrow \quad 2C_3$$
ribulose   carbon dioxide   intermediate   phosphoglyceric acid

Phosphoglyceric acid lies (see page 150) on the well-mapped pathway of glucose metabolism; some of it can be used to manufacture fructose and glucose phosphates (see page 170) whilst some of the other molecules of phosphoglyceric acid are recombined through a maze of interlocking reactions to resynthesize ribulose phosphate once more. ATP is used, finally, to rephosphorylate ribulose phosphate to ribulose diphosphate, and the cycle can start up again. Most of the enzymes concerned are precisely those of the pentose phosphate pathway and the glycolytic pathway that we have already discussed, the exception being that enzyme which actually initially fixes the CO_2, ribulose diphosphate carboxylase, which is a large complex allosteric molecule with a molecular weight of around 300,000, subject to inhibition and activation by many different substances. Needless to say, this particular enzyme is closely concerned with the control of the Calvin cycle, the metal ion Mg^{++} playing a crucial part in the control process.

But the essential point to note is that, with the exception of the apparatus responsible for the splitting of water and hence providing the primary energy source, all the reactions of photosynthesis, fixation of carbon dioxide, and synthesis of sugars follow pathways with which we are already familiar in the biochemistry of the animal cell. Once again, we find that what at first sight appeared to be a major difference in biochemical systems, between photosynthetic green plants and heterotrophic animals, is in fact more startling for its similarities.

COMPARATIVE BIOCHEMISTRY AND BIOCHEMICAL EVOLUTION

We have argued in this way without intending to deny the very real evidence of interesting biochemical distinctions between various species. And it is possible to interpret such differences in terms of a biochemical evolutionary process. Humans are probably right to feel themselves more advanced than anaerobic bacteria in that they can oxidize their food all the way to carbon dioxide and water, which demands more biochemical finesse and subtlety than the micro-organism, which can only tap off a small portion of the potential energy of glucose before being obliged to discard as refuse such energetically potent substances as alcohol or lactic acid. Similarly, very primitive organisms contain in their cells neither nucleus nor mitochondria, whilst more advanced ones have evolved them as an obvious contribution to the stability and efficiency of the cell.

One of the most interesting of such examples of biochemical evolution in action has been demonstrated by Ernest Baldwin in his study of the nitrogen-elimination mechanism. It will be remembered that we have described the problem faced by the mammal in disposing of the nitrogen produced during the breakdown of protein; this nitrogen exists in the form of ammonia, and ammonia is extremely poisonous even in very small quantities. Thus the animal, in order to avoid building up lethal quantities of ammonia, converts it instead into urea and excretes it in the urine. Not all animal species dispose of their ammonia in this manner, however. Bony fishes, for example, are content to excrete ammonia intact and without further conversions, whilst reptiles and birds instead produce not the soluble urea but the highly insoluble uric acid, which is excreted as solid nodules.

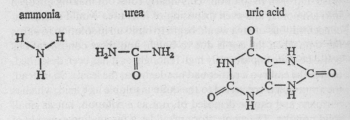

Why this difference in metabolism? Baldwin showed that it could be related to the availability of water to the different species. In fresh-water fishes, water is constantly available and continually diffuses into and out of the fish. Under these circumstances the ammonia formed will be washed out of the bloodstream as a very dilute solution and carried away into the surrounding water, without there being time for toxic concentrations to accumulate. Bony sea-fishes are in a similar position, complicated by the fact that the water in which they live contains salt at a higher concentration than that in the blood, and they therefore have to take steps against either too great an influx of these salts from the sea, or too great a loss of water into the sea. So they can only excrete *part* of their nitrogen as ammonia, but have to convert the remainder (about one-third) into other less poisonous substances.

Land animals, though, are in a quite different position. They have to *conserve* water to avoid dying of dessication. So they cannot afford to dilute ammonia down with large amounts of water before getting rid of it; instead, they turn it into the far less dangerous urea. The most interesting example of this adaptation is provided by the frog, which whilst a tadpole excretes nitrogen as ammonia, but, on changing into its adult, land-based form, also rearranges its internal biochemistry so as to make urea instead.

Finally, the birds and reptiles. The young of these species are hatched from eggs, but unlike the water-borne eggs of fish, the eggs of birds and reptiles are laid on dry land. In order to avoid complete drying out while waiting to hatch, they therefore have hard shells which are impermeable to water. But this impermeability means that they have no way at all of disposing of their waste nitrogen in solution. This clearly rules out making nitrogen into ammonia, and even disposing of it as urea would mean that concentrations of urea would begin to build up in solution towards the time when the egg is due to hatch, and these concentrations could be quite unpleasantly high (enough, as it has been described, to give the embryo a rather bad headache, at the least). So instead, the ammonia is converted to the quite insoluble uric acid, which is harmless and can be disposed of, not as a solution, but as small solid nodules. The whole story provides a fascinating example of the operation of natural selection at the biochemical level.

But despite these examples of biochemical evolution and modification in progress, to which more are being added as the science of comparative biochemistry matures, it remains true that the biochemical composition and organization of all the forms of life now present on earth demonstrate unity; a unity quite at variance with their more obvious differences in gross structure and behaviour. Most of us would hesitate to compare ourselves with fish, typhoid bacteria, cancer cells, or even oak trees, yet the fact is that we have very much more in common with them than we might have guessed.

Why? The forms of life that we know today have evolved over several thousands of millions of years, in the course of which they have branched out into such a host of diverse directions that their outward resemblances are remote. Yet during the whole of this evolution, their biochemical forms have remained remarkably constant. The only convincing explanation of this is that these biochemical parameters were established and fixed *before* the species began to evolve along the different pathways that biologists have traced. Insofar as the biochemical forms are identical, it must be the case that all existing species had a common ancestor about whose external shape and form we can only guess but about whose biochemistry we may be certain that it was very similar to the bio-

The Unity of Biochemistry

chemistry of living things today. We can express this relationship thus:

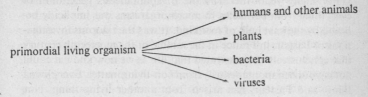

primordial living organism

humans and other animals

plants

bacteria

viruses

THE ORIGIN OF LIFE

It is such an analysis that has led biochemists into one of the most exciting of their present-day hunts – that for a convincing description of the origin of life on earth. If all current-day manifestations of life can be accounted for in strictly chemical and physical terms, as every biochemist is of course convinced, it follows that it ought to be possible also to describe the origins of these present-day life forms in chemical and physical terms as well. The alternatives would be either to assume that the primordial ancestor of both beast and human was set up by some non-chemical and physical intervention and then left to go along under its own chemical and physical steam for ever afterwards; or to maintain that life in one form or another has always existed, for as long as the universe itself, and therefore had no need to arise specifically anywhere.

Neither of these possibilities is intellectually satisfying, nor are either of them at all necessary provided we can demonstrate a convincing way in which life could *possibly* have arisen without violating chemical and physical principles. If we can provide such an account, in broad terms, it does not matter unduly if some of the details later have to be amended in the light of further scientific advances; the important thing is to show that, even in our present-day limited scientific state, we can nonetheless provide an explanation which is logically satisfactory. It is up to those who wish to mystify the nature of life then to criticize our hypotheses, if they

279

are determined to try to show that life could not have arisen in the way we propose.

For many years the principle of spontaneous generation of everything from barnacles to micro-organisms was implicitly believed by the vast bulk of mankind. It was the rigorous investigations of Pasteur in France in the middle of the nineteenth century that conclusively demonstrated that life as we now know it could not have arisen spontaneously from non-living matter. Every living thing, said Pasteur, has arisen from another living thing. Nor should this surprise us, for even the simplest of present-day living organisms are highly complex, highly improbable molecular structures, whose chance assembly from their elements would involve odds of such astronomic unlikelihood that we may regard it, for practical purposes, as impossible. The chemicals which compose present life forms require to be synthesized by specifically catalysed reactions, and these specific catalysts are themselves the product of the living organism and cannot arise spontaneously. If we are to seek for the origins of the complex of attributes that we regard as life today, we must assume that these attributes evolved only slowly over the thousands of millions of years of blank history that separate the origin of the earth from even the earliest living form whose fossilized traces we can now observe.

In order to provide some account of the way in which these attributes may have arisen, we need to be able to make plausible suggestions about the conditions that existed on earth at that remote time. The first rigorous and systematic attempt to do this was made in the 1930s by A. I. Oparin in the Soviet Union, and in its broad form his thesis is still the most satisfactory, though the advances of knowledge over the last thirty years have demanded many amendments in detail. Before Oparin, difficulties had arisen because it was assumed that the earth's primitive atmosphere was largely oxygen and that the first organisms to evolve must have been capable of performing photosynthesis in order to trap energy and synthesize the organic substances they needed. Yet photosynthesis, as we have seen, is a highly complex process clearly only possible to already well-developed and highly-skilled organisms. This dilemma was resolved when Oparin was able to point out that the atmosphere of the primitive earth, far from being oxygen-rich, must have resembled that of the other planets, containing vast quantities of hydrogen, methane (marsh-gas, CH_4), ammonia, and carbon dioxide. The present-day atmosphere has replaced this primitive one precisely *because* of many millions of years of life, turning methane and carbon dioxide into organic chemicals, and, by way of photosynthesis when it finally evolved, releasing oxygen.

We may picture the primitive earth as containing huge warm oceans in which were dissolved a variety of salts derived from rocks, and over which hung an atmosphere of gases which would rapidly be lethal to any presently living organism. Under these conditions, a number of organic compounds would have begun to be formed and scattered in solution throughout the sea. The formation of these compounds would have depended precisely on the *reducing* atmosphere and the steady influx of energy in terms of light and ultra-violet radiation from the sun, for in those circumstances CO_2, H_2O, CH_4, and NH_3 can react to give a mixture of products including amino acids, urea, and many other substances.

An interesting experimental verification of this statement has been provided by Stanley Miller in America, who passed an electric charge through a gas mixture of hydrogen, methane, and ammonia in a closed water bath for periods of twenty hours or more. At the end of this time the products were analysed and found to contain more than eight different amino acids and seven monocarboxylic and dicarboxylic acids, all of which are amongst the basic building blocks of present-day organisms. Similar experiments have even been able to demonstrate the synthesis of ATP and of small proteins under completely non-biological conditions of this type.

Thus the primitive ocean must have steadily increased in organic content. These substances would have interacted with one another to form a whole range of new substances. The surface of the rocks and clays of the beds of the shallow seas, containing iron, magnesium, and copper, would have provided catalytic surfaces on which the organic substances would have begun to collect and to polymerize. As a result, short-chain peptides and nucleic acids, and possibly carbohydrates as well, would also have begun to accumulate, both bound on to mineral surfaces and free in solution in the seas.

What followed was the most critical stage in the process. It has been known for many years that solutions containing large molecules, such as the polymers of amino acids or carbohydrates, have a remarkable tendency to break up into small droplets containing the polymers in concentrated form, leaving the surrounding water comparatively free of dissolved substances. Salts and low-molecular-weight organic substances present in the solution also tend to be sucked into these droplets together with the polymers. This phenomenon is called *coacervation*, and has a perfectly logical, though somewhat involved, explanation in physical laws. Such *coacervate drops* may be formed from mixtures containing, for example, gelatin or gum arabic, and have been extensively studied in the branch of physics known as *colloid science*.

Oparin argued that, in the primitive oceans containing polymeric organic compounds, just such coacervate drops would have begun to be formed – the organic material would all have tended to coalesce into small, highly concentrated droplets. Within the droplets, the different compounds which had collected would have begun to interact with one another because of their new proximity. In some of the coacervate drops the results of these interactions would have been to make the drops unstable – by changing the pH, for example – and they would have broken up once more. Others, though, would have remained stable for longer periods, and with the passage of time have begun to grow as they collected into themselves more chemicals. But coacervate drops have an optimum size and, if they grow beyond it, they split into two or more smaller fragments, the composition of each of which will be similar to that of the parent drop.

And so the process would have continued. Unstable coacervates would have broken down and their organic material have become available once more for incorporation into stable ones. Stable coacervates would have grown and divided. Within them, more and more complex polymers would have been formed. Metal ions acting as catalysts for favoured reactions, and coenzymes such as nucleotides, would have become more active as they became bound to the peptide polymers which were the forerunners of proteins, thus forming *proto-enzymes*. Over the course of many hundreds of millions of years, the oceans would have become peopled with these stable, reproducing, primitive, semi-living droplets. At some stage during this period in their development, the nucleic acids and proteins must have arisen as interdependent and mutually-synthesizing molecules, to form the forerunners of the DNA-RNA-protein complex which is today responsible for genetic transfer.

Meanwhile, as the coacervates continued to accumulate into themselves the organic substances in solution in the ocean, the availability of these substances must have steadily diminished. An evolutionary period must have arisen when there were not enough preformed organic molecules to go round. At this stage, natural selection would inevitably have favoured those coacervates which could make use of inorganic energy sources such as hydrogen sulphide – for a brief period, such autotrophes must have flourished more extensively than they do today. It would seem likely that it was at this stage in evolution that the mechanism of photosynthesis developed. Once the photosynthetic and autotrophic organisms had evolved, though, the conditions of existence for other organisms must have changed for the better – oxygen would have begun to appear in the atmosphere and the stock of preformed organic material in the ocean have risen sharply once more. Ultimately, a self-regulating 'carbon cycle' between heterotrophes and autotrophes would have come into play, and the era of life as we know it today would have opened.

Such, in barest outline, is the Oparin hypothesis as it has been developed over the last thirty years by biochemists in Russia and elsewhere. Obviously, it raises innumerable questions and problems. Some are complex chemical issues, such as that of the origins of the universal existence in living organisms, but not in non-living nature, of 'assymetric' molecules like those of the amino acids or sugars (see page 33). Nor are there yet very satisfactory accounts of the polymerization steps which produced proteins, nucleic acids, and so on from their more primitive ancestors.

In addition, alternative hypotheses to Oparin's place the synthesis of prebiotic molecules in the hot, dry atmosphere of early volcanoes rather than the oceans (there are some laboratory experiments showing that peptides and small proteins can be abiotically synthesized under these conditions). Some still maintain, like the astronomer Fred Hoyle, that the earth was 'seeded' with preformed macromolecules present on comets. Indeed, since the development of space probes and the possibility of seeking for life forms on other planets, a whole new area – part experiment, part theory, called *xenobiology* – has grown up around the discussion of such ideas.

Other questions are more theoretical and philosophical. Just when, for example, in this evolution of living from non-living, can we be said to have stepped across the border between the two? What, in fact, is the definition of living as opposed to non-living? There are those who try to bring to bear rigorous tests on these questions – many biochemical geneticists, for example, are convinced that a system can only be described as living when it contains a nucleic acid-protein complex capable of precise self-replication and mutation. Clearly, a key feature of today's living organisms is their capacity accurately to reproduce themselves. When and how did this evolve? Is the present genetic code an evolved form from some primordial ancestral, simpler version? Did proteins once copy each other? These questions are not *merely* scholastic; even though they are not open to clear experimental answer, they remain deeply fascinating.

However, to draw a hard and fast line between the living and the non-living is probably not a useful attempt. Clearly some things – dogs, flowers, yeast cells – are alive. Others – such as molecules of salt, urea, or amino acids – are not. Between the two extremes lies an uncertain half-world filled with coacervate drops, viruses, and some biochemical preparations like isolated mitochondria or nuclei. There is no hard and fast dividing line between living and non-living, any more than there is between a fertilized ovum in the womb and a full-grown adult, or between a raw and a hard-boiled egg. The two extremes are quite different, but the one is converted to the other by an infinite series of small steps, and it is only at the extremes that one can be very precise.

Such issues have raised heated debates at the various congresses on the origin of life that have been held over recent years. But they are debates which have been fought out within the framework laid down by physical and chemical theory and its applicability to complex systems. This is not to say that life reduces to 'mere' chemistry and physics. There are biological principles which express the organizing relationships between macromolecules, cells, organisms, and which include within them an understanding of historicity; biological systems have to be understood in temporal as well as molecular terms if their development and evolution are to become meaningful. Such principles, however, are materialistic; to understand the existence and origins of life, and of humans, needs no recourse to principles outside those of the material world.

CHAPTER 14

CAN BIOCHEMISTRY EXPLAIN
THE WORLD?

In this book, we have moved rapidly across the field covered by modern biochemistry. We have tried to do it fairly systematically, to separate the various stages in thought, experiment, and theory which characterize the biochemical approach to life, and at the same time to show how biochemists now believe that they can draw up a general balance-sheet of life which can account, in more or less broad terms, for those aspects of living behaviour we can study. Aspects of many fundamental life processes can now be described in quite precise molecular terms, analogous to those in which the chemist writes equations for the reactions of simple acids and alkalies, or the physicist for the quantum energies of the electrons of the reacting molecules. For many other qualities that go to make life we are far from being able to do this; our understanding is still too superficial. Furthermore, in many respects there are still fundamental theoretical problems in knowing what a 'complete' biochemical description of a living system would be.

There are some molecular biologists and biochemists who believe that a total description of the physics and chemistry of the cell would readily extrapolate to a total description of the organism. For example, the entire nucleotide sequence of at least one simple virus is now known. But does such a specification say all there is to say about the properties of the virus? Or are there other things to say which, as I believe, can only be specified in terms of the history of the virus as an organism, its relationship to its environment and to its host or potential host cells.

Today's molecular biologists, in their insistence on a rigid genetic, molecular reductionism, a total explanation of all there is to say in molecular terms, are, it seems to me, committing a philosophical and scientific error akin to that of the physicist Kelvin in the nineteenth century who argued that a 'complete' physics was only possible when all phenomena could be reduced to mechanical

286

analogues – clockwork models. This mechanical molecular materialism, which underlies the crudity of the 'central dogma' of molecular biology that we discussed in Chapter 10, will need to give way to a much richer understanding of the need to interpret the phenomena of life at a series of levels, from the molecular to that of the population, and that we must never see any given level as either fundamental or static. All life has a history, a biochemical as well as an evolutionary and developmental history, and the task of the biochemist becomes that of understanding living processes at just one of these levels, and of collaborating in the discovery of the translation rules that relate biochemistry on the one hand to physics and chemistry, on the other to physiology, psychology, ecology.

The account of biochemistry which has been given in these pages has perhaps falsely given the impression that this biochemistry represents the inevitable conquering march of science out of an ignorant error-ridden past into the glorious light of modern understanding. Such an impression would be quite wrong. Certainly the advance of biochemistry in the last half-century has been phenomenal, and we are still in the thick of this progress; there is no sign yet of slackening off. Biochemistry is still a young science, and many things only half-understood or mistakenly believed today will have to await a second and a third generation of researchers from now before they can be fully comprehended. Then those of us who have the temerity to publish our theories and concepts as those of the triumphant biochemistry of the 1970s will either be forgotten in the inexorable advance of science or at best half-remembered as those whose insights were later verified or disproved. If we have emphasized in this book recent results and modern work, it should not permit us to forget that all of us now working in biochemistry stand on the shoulders of the chemists and physicists of the nineteenth century, and on those of the pioneers who created biochemistry in the early 1900s as a science where none existed before. Too many graduate biochemists and postdoctoral researchers seem to work on the assumption that what wasn't published in the most recent issues of *Nature*, *Science*, or one of the biochemical weeklies is archival, and what happened before 1960 prehistoric. In the last few years only, there have been some first steps towards the creation of a history of biochemistry; more of us need to study it.

It is not only that we have to keep in mind how much is still *not* known, some of which we have deliberately hinted at in the last few chapters; it is also that experiments, facts, descriptions which seem complete in one context may, with newer and greater understanding, need to be reshuffled and reinterpreted. Biochemistry has not yet been through the convulsions of the transition from Newtonian gravitational theory to Einsteinian relativity, that shook the foundations of physics in the early part of the century; it has certainly been through a pre- and a post-Copernican phase. The disputes over energy transport and oxidative phosphorylation are probably only a foretaste of the way thinking will have to change as biochemistry moves into the last decades of the twentieth century. The mechanisms of control processes and the regulation of the cell, the intimate details of the biochemistry of cell structure, the functioning of the cell as part of the organism as a whole, the biochemical mechanisms of genetic reproduction, hormonal control, and memory, to say nothing of the application of biochemical understanding to the prevention and cure of illness, disease, and malnutrition – these are all problems which seem to us today as large and in many ways as difficult and intractable as the determination of protein structure or the elucidation of the citric acid cycle did to our biochemical elders.

All of these are soluble problems, granted support, time, and a theoretical approach which avoids an arid molecular reductionism. And such solutions are of importance not only to biochemistry itself, but outside too, for the practical application of results based on biochemical technology could be of great significance. So far, biochemistry has been associated with relatively little derived technology. Biochemists are to be found in the wine and food industries. (Pasteur was perhaps the first.) The modern pharmaceutical industry swallows the talents of others, although the biochemical rationale for the mode of action of nearly all drugs is still lacking and much of the work of devising new agents a sort of molecular roulette imposed by market demands rather than human needs. Industrial microbiology is a growing field, as more and more complex substances are being produced by allowing the enzymes within micro-organisms free rein, rather than by chemical synthesis. The proportion of biochemistry graduates who go into industry, at least in Britain, is steadily increasing.

But the breakthroughs likely in the next few years portend far more than this: potentially revolutionary medical technology; the advance of molecular biology toward the vexed field of genetic engineering; the application of understanding of brain mechanisms as techniques of mind control by precisely tailored molecules. Such possibilities have been actively canvassed. Will they be beneficial? Above all, this depends on the shape of the society which permits or denies their application. Some have their doubts. The development of anxieties over both the theoretical prospect of biological warfare and the actual use of chemical warfare do little to allay them. The recent furore over genetic engineering, in which molecular biologists themselves attempted to raise the question of the actual and potential hazards of techniques that were currently under development, has highlighted some of these problems.

It is not a question of biochemists, or molecular biologists, as gods in white coats threatening, or sensitively refraining from threatening, the future of the rest of humanity; that depends more on the structure of our society than the structure of our science, though each helps determine the other. However, biochemistry is 'special' knowledge, to which access is limited. Those who have it, have also the responsibility of spreading this knowledge, helping it to serve, not oppress the people, and learning the limitations of their own special knowledge in the process.

To answer the question which forms the title of this final chapter: biochemistry by itself alone is not enough to explain the world or even the human portion thereof. However, it is an essential part of the totality of that explanation which is the goal of true science.

A NOTE ON FURTHER READING

There are several good biochemistry and cell biology text-books available which could be read without difficulty by anyone who has got so far through this book. Amongst them we could include *Comprehensible Biochemistry* by M. Yudkin and R. Offord (Longmans, 1973) and *Biological Chemistry* and *Basic Biological Chemistry* by H. R. Mahler and E. M. Cordes (Harper and Row, 2nd ed., 1971); *Biochemistry* by A. L. Lehninger (Worth, 2nd ed., 1975) is excellent but expensive. A more general biological text is *Cell Structure and Function* by A. G. Loewy and P. Siekewitz (Holt Rinehart & Winston, 2nd ed., 1970).

Special topics are enlarged upon in, for instance, *Molecular Biology of the Gene* by J. D. Watson (Benjamin, 3rd ed., 1976), *Membranes and their Cellular Function* by J. B. Finean *et al.* (Blackwell, 1974), *Hormone Action* by A. M. Malkinson (Chapman & Hall, 1975) and *Basic Neurochemistry* by R. W. Albers, A. J. Siegel, R. Katzman and B. W. Agranoff (Little, Brown, 2nd ed., 1976). There are a number of reprints of *Scientific American* articles on cell biology and biochemistry published by Freeman, which are excellent value for money, and so too are the course units of a number of relevant Open University courses, especially S202 (*Biology: a functional approach*, Open University Press, forthcoming) and S322 (*Biochemistry and Molecular Biology*, Open University Press, 1977). These are available from university and general booksellers.

The history of biochemistry too has begun to be a topic in its own right and J. S. Fruton's *Molecules and Life* (Wiley, 1972) is well worth reading. So is *The History of Cell Respiration and Cytochrome* by D. Keilin (Cambridge, 1966) and *The Path to the Double Helix* by R. Olby (Macmillan, 1974). Finally, for the interested reader wanting to keep abreast of the most significant new developments in biochemistry and related sciences, there is no substitute for one of the popular science magazines such as *Scientific American* or *New Scientist*, which carry articles often by leading specialists in the field. These are frequently brilliant and not obtainable elsewhere and a subscription to one or both would pay off handsomely.

INDEX

291

Index

Index

Index

Index

Index

Index